Routledge Revivals

Reform of Metropolitan Governments

Originally published in 1972 and the first in a series on the governance of metropolitan regions, this study aims to explore governmental interaction with people and public interests and institutions in Metropolitan America. These papers discuss issues of how governance can be improved and the federal role in Metropolitanism as well as suggesting ways in which political reform can help. This title will be of interest to students of Environmental Economics and professionals.

Reform of Metropolitan Governments

Steven P. Erie, John J. Kirlin, Francine E. Rabinovitz, Lance Liebman and Charles M. Haar

RFF PRESS
RESOURCES FOR THE FUTURE

First published in 1972
by Resources for the Future, Inc.

This edition first published in 2016 by Routledge
2 Park Square, Milton Park, Abingdon, Oxon, OX14 4RN
and by Resources for the Future, Inc.
711 Third Avenue, New York, NY 10017

Routledge is an imprint of the Taylor & Francis Group, an informa business

© 1972 Resources for the Future, Inc.

Publisher's Note
The publisher has gone to great lengths to ensure the quality of this reprint but
points out that some imperfections in the original copies may be apparent.

The publishers would like to make it clear that the views and opinions
expressed, and language used in the book are the author's own and a reflection
of the times in which it was published. No offence is intended in this edition.

Disclaimer
The publisher has made every effort to trace copyright holders and welcomes
correspondence from those they have been unable to contact.

A Library of Congress record exists under LC control number: 71186474

ISBN 13: 978-1-138-96081-7 (hbk)
ISBN 13: 978-1-315-66014-1 (ebk)
ISBN 13: 978-1-138-96085-5 (pbk)

Reform of Metropolitan Governments

NO. 1 IN A SERIES ON

The Governance of Metropolitan Regions

LOWDON WINGO, SERIES EDITOR

Distributed by

The Johns Hopkins University Press, Baltimore and London

Reform of Metropolitan Governments

Papers by

STEVEN P. ERIE, JOHN J. KIRLIN,
and FRANCINE F. RABINOVITZ

LANCE LIEBMAN

CHARLES M. HAAR

Published by Resources for the Future, Inc.

Contents

Commission on the Governance of Metropolitan Regions

Foreword

The papers in this volume and its companions are products of a long-standing interest of Resources for the Future in the welfare and development of metropolitan America. More particularly, they stem from an RFF-sponsored project that was launched in the spring of 1970 with the convening in Washington of an informal Commission on the Governance of Metropolitan Regions. Chaired by Charles M. Haar of the Harvard Law School, it is composed of scholars, practitioners, and experienced observers of the metropolitan scene.

"Governance" in the title—The Governance of Metropolitan Regions—is meant to imply something more than government. Webster defines it as "conduct, management, or behavior; manner of life" in addition to "method or system of government or regulation." In these papers, and those that will follow, the authors are concerned not only with the apparatus and process of government in the ordinary sense but also with the total interaction among people in their public capacities and interests, and between people and the public institutions. The dominating question is: How can the governance of metropolis be improved? And next: What must we learn to achieve this? Unless early progress is made in these directions the danger that hard-pressed American cities will crack under the multiple strains of old and new problems will be very real.

RFF did not embark on this effort expecting that metropolitan political reorganization would solve all metropolitan problems, but we are inclined to think that it will help, because we have seen so many obvious steps frustrated by the way in which history has organized our urban political life. Although the reform of the institutions of metropolitan governance is hardly a sufficient condition for the solution of major urban problems, our intuition is strong that it is a necessary one.

This, then, is the theme of the RFF project—the interrelationship of metropolitan problems and governmental structure. It has formed the basis of the deliberations of the Commission and has guided the preparation of the exploratory papers published in this series. The papers do not exhaust the issues of metropolitanism; their purpose is to add some dimensions to an already rich literature, some options for the policy makers. No blueprint for the future is presented, no definitive list of recommendations; the results hoped for are breadth of view, depth of perception at several critical places, and illumination of practical alternatives for action.

Many people contributed to this effort. Charles M. Haar, as chairman, negotiated the contribution of papers and materials. Lowdon Wingo first proposed the Commission as an effective exploratory device, administered the program for RFF, and oversaw the publication of these volumes. Daniel Wm. Fessler of the University of California Law School at Davis has been a valued advisor throughout. Professors Daniel M. Holland of M.I.T. and Karl Deutsch of Harvard made valuable contributions from their time and experience. Michael F. Brewer, while Vice President of RFF, was a faithful and effective participant in the enterprise from its inception. Add to these almost twenty authors and an equal number of Commission members and you have the elements for some new insights to the metropolitan problem. Some are to be found in these papers; more are to be hoped for from succeeding phases of the project.

Joseph L. Fisher
President, Resources for the Future, Inc.

Introduction: Logic and Ideology in Metropolitan Reform

LOWDON WINGO

However convincing the logic of governmental reform, the reformers must ultimately confront the fact that radical changes in the institutions and processes of society imply extensive redistribution of societal goods—wealth, power, security, honor—and it is hardly surprising that those asked to surrender what they consider to be disproportionate shares will resist such demands and develop a counter-logic to sanctify their resistance. In the real world the confrontation of rival logics marks only one level of the conflict, the purpose of which is to reduce the neutral blocks of political power, for the uncommitted constitute uncertainties in the calculations of both sides as they approach the decisive conflict—the political engagement. "Reasoning together" can make the politics of reform more rational, less rhetorical, more realistically related to anticipated gains and losses.

Government institutions embody elaborate rules for the competitive games people play in the pursuit of societal goods. Frequently a particular set of rules results in some people losing too often or in others sensing advantages that could be gained by changes in the rules. An organized collection of such chronic losers and game-playing opportunists opting for a new game is what we call a reform movement.

It is an observable fact, however, that in most cases few are interested in changing the game: the familiar present evil is more often than not preferred to some uncertain future good. Most of the time most people will settle for a new deck or perhaps simply a new shuffle of the cards. The conditions for political reform, then, require some critical mass of the unsatisfied. Lacking this, there is no logic strong enough to bring about the transformation, and that is what the gap between Utopia and Cleveland is all about.

1

This book analyzes metropolitan governance. It is about reform and so constitutes an exercise in its logic, with propositions relating changes in government institutions to the mitigation of social problems. Nevertheless, it is not partisan—it simply seeks to penetrate the rhetoric of reform and counter-reform to find some real-world referents to the propositions advanced.

The existence of social problems does not of itself make a prima facie case for the reform of governmental institutions, notwithstanding a deep-seated political reflex to reorganize in the face of new problems. Indeed, it is not clear how the relationship between problems and an institutional reform ought to be structured. Clearly the solution to some kinds of social problems is not related to anything governmental institutions in metropolitan areas can do anything about: the concentration of poverty in metropolitan inner cities can be traced to *national* economic forces and conditions, i.e., to the inter-relations among metropolitan economies; the public health of metropolitan regions is a function of the age, sex education, income, and health history characteristics of the population, which, in an open system in the short run can be little influenced by the governmental institutions of a metropolitan region. Other problems have solutions which fall largely in the local level: the demand for participation in the social decisions which affect one's life, or the desire for amenable physical environments. In between these polar cases lie most of the problems that vex metropolitan society, problems responsive both to exogenous change and to social developments within the region. These are the problems which provide arguments for reform. They are problems characterized by "spillovers," benefits or losses accruing to interests in one jurisdiction as a result of events or activities taking place in another. We cope with spillovers by "internalizing" the consequences, and internalizing by definition involves institutional changes.

How does a logic of reform take shape? It emerges whenever "preferred" solutions to a problem are foreclosed by existing institutional arrangements. Solutions to social problems have a way of redistributing social "goods," not only among persons, but also among jurisdictions. When the losers are not good sports, they can use existing institutions to deny the benefits of the solution to the region as a whole. From their point of view, they are simply refusing to bear a disproportionate amount of the social costs (negative externalities) incurred by the solution, and in our federal system they can make it stick.

Segregation of metropolitan housing stocks is a case in point. It is implicit national policy, for example, to progressively reduce economic and social discrimination based on race or ethnic origin. A crucial barrier to this policy has been our inability to open up the suburbs to housing for minority groups, a matter of considerable moment when one considers the rate at which employment is suburbanizing in many metropolitan areas. The ideology of suburbanism posits that the development of low-income housing for disadvantaged groups in suburban jurisdictions reduces the quality of life for existing residents and requires a substantial subsidy from the rest of the local citizenry to pay the resulting incremental costs of public services which far exceed incremental revenues.

Constitutionally, the state *could* overrule the locality, but it is unlikely to do so on the simple merits of the case. The state position will depend on how many state legislators with *no* direct stake in the local problem can be captured by one side or the other (because of existing institutional arrangements districts *not* involved at all in the problem will be rewarded by the resulting competition for their legislative support). The problem might be resolved if the beneficiaries of the policy could and would compensate such communities for their welfare losses. Readily usable institutions for such transfers have never evolved in the U.S. federal system, however.

As policy fails to perform well, then, a reform logic builds up around a strategy of internalizing the problem: either provisions have to be made for compensatory transfers from gainers to losers, or gainers and losers have somehow to be brought under the same umbrella so that they can decide among themselves how the benefits of such a policy ought to be redistributed without an implicit veto being wielded by any of the parties at interest. It may be, of course, that the costs of compensation are excessive in the eyes of the beneficiaries of the policy, in which case doubt is cast on the justification of the policy.

In the end, the logic of reform seeks a congruence between the scope of political jurisdiction and the set of gainers and losers under a particular complex of policies for several reasons. First, we accept the proposition that all of those likely to be affected by a policy (or set of policies) should have some participation in its formulation: windfall benefits have been shown to have perverse effects on public allocation decisions, and the arbitrary imposition of costs on nonbeneficiaries violates our sense of fairness. Second, when all those affected participate in the decision, the losers can extract political concessions simply because they are a force to be reckoned with in the future. Third, however the internalized decision works out, it does not distort other decision-making activities in other jurisdictions. In principle, however, this kind of congruence can be defined with respect to only one problem, one public function, one policy. Is it still a useful normative construct when problems, functions, and policies are aggregated? The ideology of metropolitanism has so contended—in its deprecation of single-purpose regional policy institutions, such as the school district. Indeed, the argument goes that the proliferation of such single-purpose governmental institutions would present the metropolitan citizen with intolerable problems in realizing his preferences for public goods and publicly produced private goods. Since the consolidation of the regional fisc, the argument continues, is essential to democratic public choice, the proliferation of single-purpose agencies supported by earmarking the financial resources of the metropolitan public economy can only dissipate the ability of a region's citizenry to allocate its public resources in a manner consistent with manifest regional preferences. The conclusion made to follow from this argument is that metropolitanism needs to pursue a strategy of comprehensive internalization of metropolitan policy problems—regional general government is the obvious political structure for so doing.

Ideology aside, the policy problems of metropolitan regions have two crucial dimensions. One, noted only in passing, has to do with the relationship of a region to the rest of the world. The other and more obvious dimension is demonstrated by the fact that independent communities within densely settled urban areas experience special difficulties in their relationships with each other in pursuing common goals. Unilateral dependency relations coexist with an incentive structure that rewards the localities which can capture benefits generated by their neighbors and impose the costs of achieving their own policy objectives on others. In short, the pervasiveness of the spillover phenomenon populates the world of metropolitan communities with exploiters and victims—the former prosper, and the latter become "the urban problem." *Both* are inseparable parts of the metropolitan problem: the exploiter-victim relationship consists of spillovers which result from the fact that jurisdictional boundaries of government institutions unavoidably intersect the complex web of social and economic transactions normal to dense concentrations of population. The need to mediate these relationships constitutes a major justification for reform of metropolitan governance.

Such reform does not have to be evaluated in the abstract. A rich variety of governmental innovations at the local level has appeared in the recent past. Cities and counties have been merged. Some cities have achieved metropolitan government by annexation; others have federated. States have taken over metropolitan governmental responsibility, as in California's Bay Area Conservation and Development Commission, or created new levels of government, as in the Twin Cities Metropolitan Council. And in the 60's appeared increasing numbers of voluntary metropolitan councils of local governments (COG's) which are, in principal, confederations without articles. In short, there is an experience to be evaluated, which is the object of the first article in this book. Professors Kirlin, Erie, and Rabinovitz have carefully examined the outcomes of these experiments.

Have they worked? The answer is ambigous. They have clearly not achieved the aspirations of their staunchest proponents. They have not reduced taxes, nor eliminated fiscal disparities, nor effected any perceptible redistribution of income. They appear not to have altered the structure of political power nor to have inspired their citizens to more vigorous participation. On the other hand, they have provided an effective interface between the component localities and the federal government and have added a new dimension of credit-worthiness to local financing capabilities.

Would "next generation" reformers, learning the lessons of these experiments, have any greater hope of success? Probably, if only because aspirations and expectations are likely to be more realistic. We will be more aware that rearranging institutions alone may at best palliate strong underlying social and economic forces. Only going all the way–disowning municipal corporations in favor of a popularly elected metropolitan government–might alter these forces in time.

The resistance to "going all the way" is broad and deeply rooted. Supergovernments are viewed as potentially insensitive to the needs and wants of localities; Metro Hall could be almost as remote as the State House and concerned exclusively with matters of such general concern that the individual citizen would have gained little in his fight to control the forces that infringe on him. The bureaucratic ossification of big city governments might not seem so intolerable in contrast to a metropolitan supergovernment either badly run or dominated by rampant professionalism.

To cope with the problem, the advocates of metropolitan government have been quick to espouse neighborhood or community government as a logical complement. The Committee for Economic Development's *Reshaping Governments in Metropolitan Areas*,[1] positing the obsolescence of municipal government, argues for such a two-tiered system. Does this variant on metropolitanism offer a better hope for more effective metropolitan governance? Lance Liebman, in the second paper in this volume, thinks it might, by clarifying responsibility for public decisions, by redefining relationships between the producers and consumers of local public services, and by shaking up a political system grown moribund with age. Dangers, also, are to be noted, but on the whole it seems worth the gamble, he believes.

In recent years the metropolitan movement has grown largely through the patronage of the federal government. The source of federal interest lies in the promise that effective metropolitan political institutions could play an important role in implementing a multitude of national policies bearing heavily on urban populations. Housing, health, welfare, transportation, criminal justice, and recreation programs of the federal government all have important metropolitan dimensions which could be more efficiently realized by a metropolitan interface between localities and the federal agencies. More than this, the federal government is a critical source of policy innovations for dealing with the problem of urban growth and change. So contends Charles M. Haar in the last chapter of this book. From his experience in the administration of federal

[1] New York: CED, February 1970.

metropolitan programs under Lyndon Johnson, Haar takes a strong position on the federal government as the key metropolitan reform-monger. He argues by implication that metropolis and the "feds" need each other in confronting the entrenched political redoubts of established municipal and state government. Metro by itself, he believes, is trivial in the world of deeply vested political interests; metro in league with the federal government can become a key political structure in a Really New Federalism.

Clearly, the logic of reform for metropolitan governance is undergoing change: the nature of the problems besetting metropolitan communities has been changing. As policy challenges, failing management and the inefficient provision of publicly produced "goods" have given way before the tensions and injustices stemming from intrametropolitan disparities in poverty and wealth, in black and white populations, in environmental quality, in jobs, in public services, and in opportunities to influence the behavior of governmental institutions bearing upon one's daily life. Increasingly, federal programs are emphasizing the role of regional—and in our case, metropolitan—institutions as key elements in national public policy delivery systems. The environment of metropolitan reform of the 70's, hence, will be very different from that of the 60's. Because of this change in the content of the metropolitan problem, the prospects for reform may be better in the coming decade. Although the New Federalism of the current administration has no room for nonconventional governmental structures, in the need to reorder governmental roles among citizenry, local government, the states, and the federal government, new and stronger metropolitan institutions of governance may be the real hope, and the logical gambit for updating federalism.

1 Can Something Be Done?
Propositions on the Performance
of Metropolitan Institutions

STEVEN P. ERIE, JOHN J. KIRLIN,
and FRANCINE F. RABINOVITZ*

In the public mind there is a pervasive notion that something must be done about urban problems. The attention of analysts, however, is as much retrospective as prospective. Just how much, they ask, has been accomplished by existing governmental efforts to do something? The aim of this paper is to sort and summarize what is known about the effects of one group of efforts undertaken since World War II—the various institutional changes that involved cooperation among or consolidation of the many governmental units in metropolitan areas. These structural changes encompass city-county consolidation, federation, metropolitan special districts, urban counties, cooperative agreements between local jurisdictions within the metropolis, county-city contracts, and councils of governments. During the past 25 years these essentially local institutional alterations have greatly changed the face of governmental decision making in metropolitan areas.

*Steven P. Erie is Research Assistant, Department of Political Science; John J. Kirlin is Professor of Public Administration; and Francine F. Rabinovitz is Associate Professor, Department of Political Science, University of California at Los Angeles.

Note: This paper was prepared for a meeting on metropolitan governance sponsored by Resources for the Future, September 28, 1970. Financial and clerical assistance were provided by Resources for the Future and the Institute of Government and Public Affairs, University of California at Los Angeles. We are grateful for comments by John C. Ries, Daniel Wm. Fessler, Winston Crouch, Daniel Alesch, Gary Schwartz, Martin H. Krieger, John C. Bollens, and Charles M. Haar, who are by no means necessarily in agreement.

The first portion of this paper indicates what has been done and compares the numerous postwar metropolitan institutional changes. We describe reform efforts, contrasting them according to socioeconomic factors, proximate causes of reform, legal bases, service arrangements involved, and the authority and selection procedures for governing bodies and officials. Next, we assess the impact of metropolitan institutional changes by presenting a series of propositions that summarize the admittedly fragmentary and inconclusive research on the consequences of reform.

At first, metropolitan reformers generally agreed on their objectives; however, there has been no single definition since of high performance for metropolitan institutions. Instead, there has been a series of goals that reflect different sets of values that intergovernmental efforts are supposed to maximize.[1] In retrospect, it appears that three different aspects of the political system predominate: there has been a concern for governmental *process*, then a shift to concern for governmental *outputs*, and finally, the emergence of a set of values emphasizing the *impact* of governmental action.[2]

Process-oriented values suggest various yardsticks: economy and efficiency, rationalizing and professionalizing decision making, and access and representation for different community groups. Most of the reform literature deals with the first two, whereas the question of access and representation is raised mainly by the interracial conflict of the 1960's. During the past decade an interest in equity in the distribution of governmental outputs also emerged. Equal protection and the expanding doctrine of state action furnish a standard for metropolitan institutions different from economy and efficiency. Finally, growing interest in consumer satisfaction, emotional effects, and individual happiness—a concern for what government does *to* people—signals yet another way of assessing the performance of metropolitan institutions.

The propositions and their implications for strategy are both very tentative, reflecting the scarcity of relevant research. Although we know a great deal about the manifest purposes behind various reforms and the difficulties in seeing them through, our understanding of metropolitan governance would

[1] Kaufman (1963) has summarized the progression very well: "What is prized at one time may be derided at another; what is neglected in one period may (perhaps for that very reason) become the central concern of the next. . . . Where governmental organization is in a flux because of highly fluid social and economic conditions, as in metropolitan regions, we may anticipate a frantic search for ways of maximizing all . . . values in the instruments being forged."

[2] The three categories are derived from Sharkansky (pp. 61–80). *Process* is the manner in which actions are taken by government. Policy *outputs* represent the service levels which are affected by these actions. Policy *impacts* represent the effect which the service has on a population, the character of governmental actions against citizens if the performance is regulatory, or the response of citizens and firms to outputs.

be vastly improved by more research on the consequences of institutional changes. Our ignorance is greatest about the impact of reform on the lives and opinions of the citizens who are supposedly better governed under the new regimes. The argument that little has been done to improve urban conditions deserves to be debated on its merits rather than conceded to by default.

Description of Postwar Metropolitan Reforms: What Has Been Done?

Discussion of postwar metropolitan reform could be restricted to major areawide reorganizations such as occurred in Toronto, Miami, or Baton Rouge. However, this category omits an impressive variety of efforts to handle urban problems that spill over local boundaries. Therefore, the notion of "reformed structure" is used in a very broad and loose sense. It includes cooperative ventures ranging from service agreements between localities to councils of government, as well as federation and city-county consolidation. All involve changes in the formal relationships among local jurisdictions, whether by altering offices and authorities, contracts, agreements, or charters. Excluded, however, are cooperative agreements reached through informal or nongovernmental mechanisms (such as local professional associations of city managers or police chiefs); annexations; school districts; and decentralizing or neighborhood-control policies. Ultimately a balance sheet on institutional reform should compare the costs and benefits of the major classes of reform—annexation, decentralization, and metropolitan integration.

To outline the extent to which the subset of metropolitan integrative reforms changed relationships among local jurisdictions, Table 1 presents the various types of metropolitan reforms. The major areawide reforms are considered individually, but for certain types of reforms—urban counties, metropolitan special districts, intermunicipal agreements, and county-city contracts—only examples are given. We do not know how typical these illustrations are, because many structural changes (such as those involving special districts) are difficult even to identify (Bollens, 1956). Similarly, it is not possible at present to document from existing studies the experience of the more than a hundred areas that have councils of government.

Given these restrictions, the postwar reforms can be described initially by differences in their legal capacity for action and in their mode of future alteration. Two major types of restructuring can be distinguished by the relative ease of dissolution—"permanent" and "cooperative" arrangements. Permanent arrangements include abolition mechanisms partaking of a more extraordinary character than intermunicipal agreements; for example, city-county contracts and councils of government require a vote of the electorate or new state legislation. In cooperative institutions disassociation usually requires only unilateral action of a member's governing board—city

TABLE 1. Postwar Metropolitan Reforms

	Area (sq. mi.)	Population 1960	Percent urban	Percent non-white	Median income ($)
I. PERMANENT[a]					
A. *Unitary* (4)					
Baton Rouge, La. (1947)	462	243,396	85.1	31.8	5,830
Nashville-Davidson, Tenn. (1962)	527	415,000	87.7	19.2	5,332
Jacksonville-Duval, Fla. (1967)	827	455,411	85.2	23.4	5,345
Indianapolis-Marion, Ind. (1970)	402	800,000	91.2	14.4	6,609
B. *Federal* (2)					
Miami-Dade, Fla. (1957)	2,054	935,047	95.6	14.9	5,348
Toronto, Canada (1954)	241	1,200,000			
Urban counties, e.g., Nassau County, N.Y.	300	1,300,171	99.7	3.2	8,515
Metropolitan special districts, e.g., Chicago Transit Authority (1947)	730	5,000,000	99.0	17.3	7,287
II. COOPERATIVE					
A. Intermunicipal agreements, e.g., Philadelphia, Pa.	2,180	3,591,400	44–100 (by county)	1.9–26.7	5,782–7,632
B. County-city contracts, e.g., Los Angeles, Calif., "Lakewood Plan" (1954)	4,060	6,038,771	98.8	9.7	7,046

dian cation ears)	No. of general-purpose govts. affected	Proximate cause of reform	Legal basis/ adopted by	Differential tax & service zones
1.9	4	Inadequate services to city fringe	Const. amendment single majority	Yes
0.3	8		Dual majority	Yes
0.8	2	Declining core city; poor services, fringe		Yes
1.4			State statute	
1.5	27		Const. amendment single majority	In effect: yes; services pro-
	13	Fiscal crises; service inadequate	Statute of provincial legislature	vided at metro level undifferenti-
2.2	70	Urbanization of unincorporated areas	Home rule statute	ated; but cities still provide some services
0.7		Provide better mass transit for metro area, service outlying sector		. . . .
-12.1	238	Interjurisdictional spillovers; inability of small cities to provide certain services	State enabling leg.; contracts among units	. . . .
2.1	78	Pressure against L.A. County providing urban services to unincorporated areas	State enabling leg.; contracts between County and cities	. . . .

TABLE 1. Postwar Metropolitan Reforms (Continued)

	Area (sq. mi.)	Population 1960	Percent urban	Percent non-white	Median income ($)
II. COOPERATIVE (Continued)					
C. Councils of Government (100+)[b] e.g., Los Angeles SCAG (1965)	38,285	7,824,000	62–98.8 (by county)	1.4–9.7	5,507–7,219
e.g., Metropolitan Washington MWCOG (1957), D.C., Va., Md.	2,345	2,076,610	12–100 (counties and D.C.)	3.9–54.8	4,460–9,317

Note: Ellipsis indicates "not applicable."

[a]Other examples are Hampton, Va. (1952), Virginia Beach, Va. (1962), and South Norfolk, Va. (1962), Carson City, Nev. (1969), Juneau, Alaska (1970), and Columbus,

council or county supervisorial unit. This is accomplished through either periodic review, as in the case of the Lakewood Plan in Los Angeles, or some form of advance notice before withdrawing from membership in a council of government. There are two forms of permanent arrangements—unitary, or one-tier, and federal, or two-level, systems. The four unitary cases in Table 1 are all examples of city-county consolidation. Federal systems, as in Toronto, Miami, and some urban counties, preserve separated authority, fiscal powers, and personnel among previously existing governments. Metropolitan special districts, such as the Chicago Transit Authority, which also maintain independent status from other local jurisdictions are special instances of federated or overlapping forms. In terms of the scope of authority, unitary systems are not necessarily more powerful than federal systems. A bifurcated legal structure, as in Toronto, can exhibit a broad delegation of authority to the metropolitan tier.

Two kinds of cooperative arrangements can be distinguished (Table 2). Some deal primarily with providing services (intermunicipal agreements and county-city contracts). Others provide a forum for consultation among local government officials. The service agreements seem to vary according to whether they are between municipalities or between cities and counties. The former are most likely among contiguous cities and deal with matters such as police radio and mutual assistance, garbage and refuse collection, or joint library use (Dye *et al.*, 1963; Marando, 1968). Contracts between cities and counties, on the other hand, contain broader and more standardized service

Median education (years)	No. of general-purpose govts. affected	Proximate cause of reform	Legal basis/ adopted by	Differential tax & service zones
9–12.2	101 (6 counties and 95 out of 145 municipalities)	Federal funding; fear of state intervention	State Joint Exercise of Power Act	. . . .
8.8–12.8	13	No real crisis; effort of one member of D.C. governing board	Corporate charter under D.C. laws	. . . .

Ga. (1970). (Burton.)

*b*If both COG's and Regional Planning Associations are included these now exceed 500.

arrangements; for example, the county may provide all law enforcement services for a municipality. Councils of government are in a separate category because they usually function as forums for consultation with little authority to provide services. Some exercise a review power over municipal applications for federal money, although they rarely appear to veto the projects of members.

The materials in Table 1 on the socioeconomic context of various arrangements and apparent proximate causes of reforms, in Table 2 on the range of service arrangements, and in Tables 1–3 on the structure of new governing bodies are meant to be largely illustrative of "what is." We can at this point make relatively few summary statements. Looking at the range of socioeconomic characteristics—area, population, percent urban, percent nonwhite, median income, education—the greatest difference seems to be region. Permanent unitary or federal systems are concentrated in the South, whereas cooperative arrangements (excluding councils of government), appear more frequently in the nation's highly populated states; for example, California, New York, Illinois, and Pennsylvania. In reformed southern metropolitan areas fewer governmental units are involved, and the areas have smaller populations with lower median incomes and somewhat larger percentages of nonwhite citizens. In Baton Rouge, Nashville, Jacksonville, and Miami, voter approval (generally an areawide majority) created the new regime, and the rhetoric of reform concerned service inadequacies in the urban fringe, not the central city. The limited analyses of southern reforms suggest race was not at issue

TABLE 2. Services Provided by Metropolitan Institutions

	Public Safety	Public works	Park, recreation	Planning	Education
I. PERMANENT					
A. *Unitary* (4)					
Baton Rouge, La. (1949)	Urban zone: city-parish rural zone: parish or municipalities	City-parish	City-parish	City-parish	State
Nashville-Davidson, Tenn. (1962)	City-county	City-county	City-county	City-county	City-county
Jacksonville-Duval, Fla. (1967)	City-county	City-county	City-county		City-county
Indianapolis-Marion, Ind. (1969)		City-county			
B. *Federal* (2)					
Miami-Dade, Fla. (1957)	Cities (records, committees, training by county)	County	County	County	
Toronto, Canada (1954)	Metro	Metro	Metro	Metro	Metro financing; municipal control & additional financing
Urban counties, e.g., Nassau County, N.Y.	Police services to 80% of county residents in special police service zone	County does water, most highways	County establishes & operates park system	County plans committees with weak powers	County jr. college; school districts
Metropolitan special districts, e.g., Chicago Transit Authority (1947)	....	....	....	....	....

14

Health & welfare	Zoning	Codes & licensing	Garbage & refuse	Public housing	Public transportation
Health: city-parish welfare; state	City-parish	City-parish	Municipalities		City-parish
City-county	City-county	City-county	City-county	City-county	City-county
City-county			Urban zone: city-county		City-county
					City-county
County	County	County		County	County
Metro	Municipalities	Metro		Metro	Metro
Countywide health and welfare	Towns & cities	Towns & cities	Towns & cities	No county activity	No county activity
. . . .	. . . .	. . . .	. . . .	. . . .	Transit authority

TABLE 2. Services Provided by Metropolitan Institutions (Continued)

	Public Safety	Public works	Park, recreation	Planning	Education
II. COOPERATIVE					
A. Intermunicipal agreements, e.g., Philadelphia, Pa.	108 police radio & 25 mutual assistance agreements	24 agreements on roads & bridges; 36 on water	No agreements	46 agreements	193 agreements on schools; 37 on libraries
B. County-city contracts, e.g., Los Angeles, Calif., "Lakewood Plan" (1954)	29 cities contract for Sheriff's law enforcement services	County provides 24 cities most of their public works	5 cities contract recreation service	21 cities contract for planning & zoning	....
C. Councils of Government (100+), e.g., Los Angeles SCAG (1965)	No committee or major activity	No committee	Parks & recreation committee	Regional planning committee: open spaces committee	No committee
e.g., Metropolitan Washington MWCOG (1957), D.C., Va., Md.	Public safety committee	No committee	No committee	Land-use committee	No committee

Note: Ellipsis indicates "not applicable."

(Harvard and Corty, 1964; Glendening, 1967; Grant, 1965, 1969), although it was popularly thought that voter support in Nashville and Jacksonville was inspired by fear that blacks would take over within the cities (CED, 1970).

As to the proximate causes of reforms, it is quite clear that most reforms are reactive; they are responses to fiscal and service crises, not anticipations of future needs. Even the Councils of Governments (COG's) may well fit this

Health & welfare	Zoning	Codes & licensing	Garbage & refuse	Public housing	Public transportation
5 agreements on milk control	None	None	33 on refuse; 84 on sewage	None	None
County provides health services to 75 cities; welfare to 77	21 cities	Sheriff does business code enforcement for 21 cities	None	None	None
Air pollution committee	No committee	No committee	Water & waste management committee	No committee	Aviation & airports committee; transportation committee
Pollution & water supply committee; health & welfare committee	No committee	No committee	Pollution & water supply	No committee	Transportation committee

pattern. The Southern California Association of Governments was a direct response to the fear of state intervention in the field of regional planning, and other COG's are creatures of federal planning requirements (Douglas, 1968; Erie, 1968). In addition there is no uniform legal basis for reorganization although some form of state authorization nearly always is required. Many do not even require voter approval. Rarely, however, do reforms rest on extra-

TABLE 3. Metropolitan Governing Arrangements

	Legislative Body			Executive Officer		
	Size	How selected	Concurrent officeholders	Official title	How selected	Authority to appoint department heads
I. PERMANENT						
A. Unitary						
Baton Rouge, La. (1949)	9	7 from City of Baton Rouge; 2 from rural districts	Most; the 7 from City of Baton Rouge	Mayor-President	Elected at large	Some; without confirmation
Nashville-Davidson, Tenn. (1962)	41	6 at large; 35 from districts	No	County Mayor	Elected at large	Nearly all; without confirmation
Jacksonville-Duval, Fla. (1967)	19	5 at large; 14 from districts	No	Mayor	Elected at large	Most; with confirmation
Indianapolis-Marion, Ind. (1969)	29			Mayor		
B. Federal						
Miami-Dade, Fla. (1957)	9	8 at large, but with district residential requirements; 1 mayor	No	Mayor	At large	No indep. power; largely ceremonial
				Manager	Manager appointed by county commissioners	Some
Toronto, Canada (1954)	33, 11 Executive Committees	Municipal aldermen with highest votes; Executive Committee chosen by vote of	Yes	Chairman	Appointed by Executive Committee; need not be elected official	Makes nominations to Executive Committee

	No.	District elections	All except supervisor from City of Long Beach	County Executive	Elected at large	Most; with confirmation
Urban counties, e.g., Nassau, N.Y.	6	District elections	All except supervisor from City of Long Beach		Elected at large	Most; with confirmation
Metropolitan special districts, e.g., Chicago Transit Authority (1947)	7	Appointed, 4 by Chicago mayor, 3 by governor				
II. COOPERATIVE						
A. Intermunicipal agreements, e.g., Philadelphia, Pa.		…		…	…	
B. County-city contracts, e.g., Los Angeles, Calif., "Lakewood Plan" (1954)		…		…	…	
C. Councils of Government (100+) e.g., Los Angeles, SCAG (1965)	103 / 15 Executive Committees	Each municipality chooses 1 from council or mayor; county from board members. Executive Committee by city & county caucuses	Yes	Chairman	By vote of Executive Committee	Largely ceremonial

Table 3. Metropolitan Governing Arrangements (Continued)

	Legislative Body			Executive Officer		
	Size	How selected	Concurrent officeholders	Official title	How selected	Authority to appoint department heads
Washington, D.C.: MWCOG (1957)	16 7 Steering Committees	1 from each county & city. Steering Committee: 1 from each jurisdiction with more than 100,000 pop.; 1 for all other jurisdictions	Yes	Chairman		

Note: Ellipsis indicates "not applicable."

governmental bases: the corporate charter of the Metropolitan Washington Council of Government is highly atypical.

Responsibilities for service provision are not consistently assigned to particular institutions (Table 2). Police service, for example, is a metro function in Nashville; Miami municipalities provide basic police services; and Dade County provides some common ancillary functions (e.g., records and training). Some patterns do exist, however. In cooperative systems, the focus is on physical and tangible activities—parks and recreation, waste disposal, or transportation systems—and not on social or welfare functions such as public housing. Similarly, the major metropolitan special districts function in the fields of water provision, sewage and waste disposal, transportation, and parks (Bollens, 1956).

Few patterns are found among metropolitan reforms in either the authority or selection of governing bodies and officials. Even in the permanent systems there is little similarity in size of governing council or the authority of the chief executive officer, although most governing bodies of the major areawide structures are elected by districts and the executive officer is elected at large. The governing bodies of metropolitan special districts are usually appointed by governors or county officials. COG's normally elect a governing body or executive committee from officials of the local member jurisdictions.

If "reform structure" means the rearrangement of areawide officers and powers, metropolitan reforms can be considered major changes. But is this a sign of change in the way things get done? An offices-and-powers concept of structure assumes that predictable patterns of behavior flow from institutional design, for the essence of the future state of affairs is thought to inhere in the new design. The question, of course, is whether patterns of decision making, policy outputs, or environmental impacts change significantly as a result of new institutional arrangements.

Analysis of Postwar Metropolitan Reforms: What Has Been Accomplished?

In assessing the performance of metropolitan institutions, the definition of performance is critical. We have suggested three appropriate sets of goals dealing with process, output, and impact effects. Advocates and analysts of metropolitan reforms have rarely adhered to any single one, and they differ as to which institution of metropolitan reform should achieve, or has achieved, particular objectives. We shall use the three categories to organize the evidence of results of metropolitan reforms. To create a balance sheet for institutional performance we shall attempt to assess the record in each of these areas, as well as the relationships among them. Remember that all these characterizations are relative in some sense to the conditions in nonmetropolitanized areas, for which the implicit baselines are not specified or necessarily clear.

Process Effects

Chronologically, process objectives were among the first goals reformers sought. These goals included economy and efficiency, professionalization of governmental decision making, and the capturing of spillovers—or externalities—of governments in a given region. In the 1960's, another type of process goal gained primary attention. The desire to incorporate previously excluded groups gave rise to the notion that the political process should also be judged by its breadth of access.

What has been the record of reformed institutions in meeting this set of goals? Our analysis of existing studies of the reformed institutions leads us to the following propositions:

Proposition 1: Although the conditions under which economies of scale might be realized are present in many reformed metropolitan institutions, there is little evidence that they are being realized in any but the most routine kinds of public services.[3]

This proposition contradicts the general belief that scale economies will take place across the board. The conventional argument is made by Dan Grant, discussing Nashville Metro:

> In the opinion of this writer metro has already eliminated many duplications and has some economies to show for this . . . . It is therefore possible to contend . . . that metro has eliminated the wasteful duplication of street maintenance and traffic engineering equipment in city and county departments . . . (1965, p. 40).

Besides eliminating duplication, more broadly based organizations are said to have had inherent scale economies, as in the incentives for cities to contract with Los Angeles County for police protection.

> Small police forces necessarily lack the ancillary services and specialized operations which make the difference between inadequate and superior police protection. The use of the sheriff's department on a contract basis leads to services being provided by an organization that is big enough to support them well and one able to provide ancillary services, including laboratories, training academies, community relations offices, youth

[3] We are grateful to Daniel Alesch for discussions on this proposition. The use of the terms "public" and "private" goods follows the definition stated by Olson: "A common, collective, or public good is . . . defined as any good such that, if any person X_1 in a group, . . . X_i, . . . X_n consumes it, it cannot feasibly be withheld from the others in that group. . . . The basic and most elementary goods or services provided by government, like defense and police protection, and the system of law and order generally, are such that they go to everyone or practically everyone in the nation" (Olson, 1968, p. 14).

services, a juvenile bureau, criminal intelligence, specialized vice squad, and narcotics squad. More effective use of vehicles and other equipment, superior records and communications systems, and a larger pool of manpower to call upon to fill fluctuating demands for service are some of the advantages associated with a more broadly-based police department (Shoup and Rosett, 1969, p. 35).

However, this impression is not supported by the available evidence on the relationship of size to scale economies, summarized in the work of Hirsch (1967), Shapiro (1963), and Dahl (1967). The most complete summary is furnished by Hirsch, who examines research on the average unit cost function of two different types of services: horizontally integrated (e.g., police, fire, education) and vertically integrated (e.g., gas, electricity, sewage) (Hirsch, 1967, Table 46). Hirsch indicates that, for private goods, increasing size of population tends to be associated with decreasing average unit cost, suggesting the presence of scale economies. With public goods, however, the cost function plotted against size is curvilinear. In populations up to 15,000 the cost function decreases. In the range 15,000 to 100,000 it remains unchanged. Above 100,000 it appears, depending on the study in question, to continue unchanged or increase slightly. Other analysts suggest that even this categorization is faulty, because the long-run average cost curve for each kind of service is different from that for the next.

Why has no one clearly identified economies associated with the provision of public goods by metropolitan institutions? One answer is that little work has been done on the subject. Another possibility is that the existing evidence on scale economies contains a bias that prevents recognition of quality gains with increased scale because it focuses solely on additions of units of service. A more common argument, which assumes service quantity as a measure, is that many functions are not subject to scale economies since they are labor intensive and not performed with new technologies in larger areas but only on a greater scale. Hirsch (1967, p. 48) states:

> It can be concluded that most governmental services (of a "public" nature) require relatively close geographical proximity of service units to service recipients; this prevents the establishment of huge primary schools, fire houses, police stations, or libraries. Urban government services are also more labor intensive, with wages and salaries often accounting for more than two-thirds of the current costs. The resulting concentration of manpower can increase the bargaining power of labor and this, in turn, increases costs. While there are some economies resulting from bulk purchases of supplies and equipment, such savings can be outweighed by inefficiencies resulting from top-heavy administration and the ills of political patronage in very large governments. Therefore, in terms of economies of scale, governments serving from 50,000 to 100,000 urbanites might be most efficient.

Still another argument deals a priori with the impact of areawide metropolitan government upon scale economies.

> The conditions which help private industry to benefit from scale economies—lower factor costs, larger and more efficient plants, and induced circular and vertical integration—often do not appear to exist when local urban governments grow or consolidate (Hirsch, 1967, p. 38).

Compared with the diseconomies associated with public goods, substantial economies of scale appear in the provision of the most common private, or vertically integrated, goods. The metropolitan special district is the institutional instrument often used to capture such economies. Yet, even allowing for economies in providing private goods, the overall picture is one of noneconomies or diseconomies, because private goods consume only a small portion of the total metropolitan fiscal outlay. Dahl (1967, p. 966) cogently summarizes the minor role played by private goods economies:

> The few items on which increasing size does lead to decreasing unit costs, such as water and sewerage, are too small a proportion of total . . . outlays to lead to significant economies; and even these reductions are probably offset by rising costs for other services, such as police protection.

Proposition 2: Reformed metropolitan institutions strengthen the influence of professionals on public policy making, particularly in physical, tangible policy scopes, such as capital projects and transportation, of a technical and noncontroversial nature.

The increased influence of professionals can be found throughout the entire range of reforms. In Toronto Metro, for example, professional staffs became the most important group initiating proposals for the Metro Council, although the chairman evaluated the political feasibility of proposals and managed attempts at passage by the council (Kaplan, 1967, pp. 69-74). Similarly, in Nashville Metro, Grant found that nearly all departments experiencing consolidation in the city-county merger sought outside technical experts and consultants to aid in the transition and tried to recruit qualified, nationally known professionals to head the various agencies (Grant, 1965, p. 41). At the other end of the spectrum, professional staffs appear to play a critical "agenda setting" function in councils of government due to their expertise and the fact that members of COG executive committees do not devote much time to doing their "homework" between meetings (Erie, 1968).

The impact of professionals appears to be greatest in those areas in which reformed arrangements have been most active: capital projects, tangible and physically oriented rather than softer, socially oriented functions. To an extent, this is a consequence of the complexity of these technical areas. Technical expertise is deferred to because public policy makers recognize their own limitations. As subsequent propositions make clear, policy makers restrict their purview largely to the *distribution* of such projects and programs, ensuring some type of parity among participating localities. Among the policy scopes in which professionals are most active (and elected officials least active) are the following: water provision, sewage disposal, capital construction projects, and transportation. Professionals in the softer areas of public policy, such as general planning or welfare, appear to exercise relatively less influence, although this too may be increased as a consequence of metropolitan reform.

Proposition 3: Federal and cooperative metropolitan governmental institutions provide a forum which facilitates diffusion of information among participating governmental units.

This benefit is most applicable to the federated form of metropolitan institutions and is the benefit most widely claimed for the COG's. "Of the many activities in which councils of government have engaged, by far the most significant is intergovernmental communications. . . . The councils have brought officials from their isolated councils and boards together socially and politically to share concerns," says one observer (Hanson, 1966, p. 6). In a 1966 survey, between 89 percent and 93 percent of the members of the executive bodies of COG's and from 63 percent to 81 percent of the general deliberative bodies of COG's felt that informal consultation was the chief accomplishment (Thomas, 1967, Table 92).

The impact of the informal process is of course reduced by its very informality. Participating units change their behavior only insofar as it is in their self-defined best interest to do so. Moreover, institutions other than areawide governmental bodies also facilitate informal interchange. This is a particularly important role for professional groups of managers.

Proposition 4: Representatives to metropolitan areawide decision-making bodies are delegates of subareas of the region, whose interests are of primary importance to them.

The tendency of representatives to areawide governing bodies to view themselves and act as delegates of specific subareas can be found throughout the entire range of reforms. However, we have no evidence concerning governing boards that are elected at large or appointed by governors, cases in which representatives might be expected to serve the interests of larger areas. Kaplan found in Toronto that representatives of Toronto Metro Council " . . . saw themselves as municipal spokesmen attending an international conference. They were municipal officials first and Metro councillors second (Kaplan,

1967, p. 83; see also p. 215). Hanson also argues that councils of government imposed self-limitations on their activities, largely explicable in terms of the representatives' persistent views of themselves as delegates of subunits (Hanson, 1966, p. 13). In the case of the New York Metropolitan Regional Council, "the members represent towns and cities according to geographical boundaries and tend to examine proposals in terms of costs and benefits to their own communities (Aron, 1969, p. 122). In a survey of COG members, respondents said *they* could view proposals from a regional viewpoint but that decision makers in the units they represented viewed proposals from the viewpoint of the units. This recognition may temper the proposals made by the representatives themselves (Thomas, 1967, p. 537).

A factor contributing to the delegate orientation is the mode of selection found in these schemes. In most major reforms—city-county consolidation and federation—representatives are elected on a district basis (Miami is the exception, with at-large elections but district residence requirements), with most municipalities retaining their independent legal status. (Only Toronto Metro experienced a consolidation of preexisting cities and that action was performed by provincial legislation.) As others have found when examining the impact of the district mode of election on councilmen's decision making in large cities (Erie, 1967), this recruitment method contains a host of incentives that favor the delegate outlook. Incentives range from the creation of a localized political organization to ensure reelection (hence deference to locally based groups and interests) to the common practice of citizens to pressure *their* representatives to intercede with bureaucracies on their behalf.

Similarly, the basic units represented in nearly all councils of government are the municipality and county. To a large extent councils are congresses, whose governing members are themselves elected officials of subunits. Kaplan's observations that Toronto representatives act first as municipal spokesmen would appear to hold here, too. However, interviews with the Los Angeles COG Executive Committee show that the body is, in contrast to Proposition 4, composed largely of *qualified* regionalists who share a concern for areawide physical and spatial problems, i.e., smog and transportation, yet view the organization as a mechanism for strengthening local home rule (Erie, 1968, pp. 22 ff.). The governing body has a distinctly suburban point of view concerning metropolitan problems. Emphasis on physical problems (see Proposition 9) encourages this orientation; we believe that the more typical delegate (or watchdog) outlook might rapidly appear if social or wealth redistribution problems were encountered.

In retrospect this outcome seems predictable. Yet reformers expected at first that joint action would induce greater mutual acquaintance and understanding, which in turn would encourage an areawide frame of reference (Jacob and Teune, 1964).

Proposition 5: In the process of decision making, the governing bodies of reformed metropolitan institutions seek unanimity and parity in the distribution of expenditures. As a consequence, capital projects are apportioned on the basis of district.

Decision makers and citizens alike view metropolitan institutions as comprised of units maintaining separate existence (as noted in Proposition 4, representatives elected at large or appointed by governors may be exceptions). The result is that decision-making norms are designed to maintain the continued operation of the areawide institution by not threatening the subunits. Parity distribution makes participants feel equitably treated. Moreover, a unanimity norm gives each decision maker a veto by which he can kill any proposal he feels is not in the interests of his subunit, irrespective of the council's voting arrangement (Hanson, 1966, p. 20; Kaplan, 1967, p. 93).

The parity norm appears strongly in the case of *capital* projects. For example, funds for streets and roads in Nashville-Davidson County are still apportioned on the basis of councilmen's districts (McArthur, 1967, p. 116). Some observers believe that parity is not only a norm but is in fact necessary to the continued existence of a metropolitan unit (Kaplan, 1967, pp. 61; 85-89; 109; 112). With parity, we would expect *service* inequities to disappear. Proposition 14 indicates that a set of political forces operates, however, to maintain service disparities among metropolitan subunits; service distribution remains relatively unchanged after reform.

Proposition 6: After metropolitan institutions have been reformed, the greatest conflict in them is generated by proposals to rearrange the institutions of metropolitan governance again.

A striking feature of new metropolitan institutions is the frequency with which attempts are made to rearrange things again and the depth of conflict this generates. In Miami-Dade County, 15 such referenda have occurred. Metro expansion forces won in all but one case, an attempt to change the office of County Sheriff from appointive to elective (Glendening, 1967, pp. 115-17). Despite the number of attempted reforms in Toronto, there was no evidence of support for abolition of Metro; only an abundance of suggested decreases in its power (Kaplan, 1967, pp. 41-64; 83-84; Smallwood, 1963, pp. 5-9; 39). Experience with the COG's is similar, although the conflict is less frequent and intense. There often have been great difficulties in creating executive or steering bodies for COG's and thereby settling the problem of representing the large number of participating units (Hanson, 1966, pp. 16-20; Thomas, 1967, pp. 220-346). In the New York Metropolitan Regional Council, for example, the continuing debate over organizational form consumed so much energy that substantive programs were slighted (Aron, 1969, pp. 144-71).

Proposition 7: Reformed institutions do not, in the short run, alter the distribution of power in the metropolitan area. Unless previously inattentive groups are specifically made aware of the differential advantages or deprivation inherent in institutional change, there will be little penetration from these groups. The norm is rather vigorous competition among elites seeking relative advantage in the new institutions.

This proposition is suggested by analyses of reform campaigns; there is no specific evidence of which we are aware from studies of Metro in operation. In the campaigns, reformers usually reach attentive and educated voters rather than relatively unsophisticated ones. Furthermore, the plan is normally so vague that inattentive people are not sure what will happen to functions about which they may in fact have a special concern. One suspects that special efforts to reach particular groups—like the Office of Economic Opportunity (OEO) legal services program—will be required to activate the normally inattentive.

Proposition 8: Reform schemes tend to guarantee that black influence remains unchanged. This reduces or arrests the rate of increase that would be experienced in the core city on the basis of black population increases. The net effect is to "freeze" black electoral power at its preadoption level. Thus, although the short-term effect is "no dilution," the long-term effect of reform may be to neutralize growing black electoral strength.

Although observers of Nashville and Miami (Grant, 1965; Glendening, 1967) indicate that Negro influence has not increased under reform, they are prepared to argue that it has not decreased either. Recent data on the estimated population (1968) in Metro Nashville's 35 councilmen's districts lend credence to this weak conclusion. Nonwhites comprise 19 percent of the areawide population and are a numerical majority (ranging from 55 percent to 98 percent) in 6 of the 35 districts, or 17 percent (Planning Commission of Nashville-Davidson County, 1968, p. 22). Grant says the districting scheme is " . . . a kind of 'racial gerrymandering in reverse,' which virtually guarantee[s] the election of several Negroes to the proposed metropolitan council" (Grant, 1965, p. 51).

But what would have happened in Nashville in the absence of consolidation? Negroes in 1960 comprised 38 percent of the Nashville population. Planning commission data show that Negroes represent 28 percent of the population in the urban services zone, which includes the former city of Nashville and the surrounding urban fringe area. This suggests that if there had been no reform or if reform had been limited to the area of dense population in and around Nashville, Negroes could have expected to control between 25 percent and 40 percent of the councilmen's seats in these arrangements. The rapid increase in nonwhite population in the core city has not been translated into a proportional increase in electoral strength.

Voting strength, however, is only one measure of political influence.

> . . . vocal black leaders have not been hesitant in expressing their dissatisfaction with reformed governments. In Nashville, charges have been made that housing, highways, and police services are not responsive to minorities and that minority political power has been diluted. Julian Bond of Georgia says black power has been diminished in both Nashville and Miami.[4]

It appears that Negro leaders and minority voters in general in reformed areas *feel* that their power and influence suffer dilution in metro reforms. Marshall concludes that

> Their predominant opinion is that in the sometimes purposeful, but most often willy-nilly shifting of governmental centers, of redistribution of resources, and of political power plays, the man of color or cultural differences tends to be the loser.[5]

The Nashville consolidation scheme apparently operates in a manner to neutralize growing black electoral strength in the core area, furnishing reason for "the charge [by black leaders] . . . that the consolidated government has diluted Negro political power" (Bollens and Schmandt, 1970, p. 306).

This proposition rests on very limited data, however. Negroes complain about unreformed government as well as reformed, and we have no studies that measure the relative weight of such opinions.

Output Effects

In the rhetoric of metropolitan reform, output is believed to be predictably and closely linked to process. Thus, services to citizens were expected to improve as a function of anticipated process effects. But, as we have already suggested, the outputs of metropolitan institutions could also be seen in a different and broader perspective, for the yardstick in the early 1960's moved from concern with volume of services to their distribution. The output-related consequences of metropolitan reforms evident in existing studies are stated in the next five propositions.

Proposition 9: Reformed metropolitan institutions emphasize the provision of tangible, private goods, especially those for which unit costs can be determined.

The evidence on this proposition is extensive. It seems to hold for perpetual institutions (Smallwood, 1963, pp. 35, 38; Kaplan, 1967, pp. 120, 126; Glendening, 1967, pp. 159-63), such as the metropolitan arrangements of

---

[4] See Dale Rogers Marshall, "Metropolitan Government: Views of Minorities," in *Minority Perspectives*, no. 2 in a series on The Governance of Metropolitan Regions (Washington, D.C.: Resources for the Future, 1972).

[5] See Marshall; quote is from Rendon, p. 9.

Toronto and Dade County, for the cooperative agreements among municipalities in metropolitan areas (Dye *et al.*, 1963; Marando, 1968), and for the Contract Services Plan (Lakewood Plan) of Los Angeles County. The policy and planning agendas of councils of government share a similar orientation, with the exception of Washington, which has unique relations with federal departments (Thomas, 1967, pp. 579-80). The New York Council initiated and then abandoned studies in the area of housing and redevelopment. The Southern California Association of Governments does not even have committees on poverty, housing, employment, or welfare.

Proposition 10: If services are transferred from a local to an areawide basis, average service levels rise. The net effect of restructuring is a per capita increase in service costs due to the increase in average service level.

This situation seems to occur in two different ways, but it appears only where areas previously receiving disparate levels of services are joined in a service area. In some cases, there is a tendency to set the new areawide norm by the subunit with the highest level of services. Sometimes linked to establishment of a professional bureaucracy for the function, this situation creates a constituency for increased services (Kaplan, 1967, p. 114). In other cases, services will be spread thin initially to achieve uniformity, generating a drop in performance. The drop spurs attempts to increase services and, after a lag of two or three years, to increase personnel and equipment (McArthur, 1967, pp. 103-15).

Rising service levels have been found in Nashville (Grant, 1965, p. 40) and Toronto (Cook, 1968; Kaplan, 1967, p. 114). On the other hand, researchers looking at the effects of Los Angeles County's Contract Services Plan (independent cities contract with the county for selected services) have found higher service levels *without* a per capita increase in service costs. This exception is attributable to the county's ability to underprice certain services such as police protection (Shoup and Rosett, 1969). The county does not charge contract cities for ancillary services, preferring to finance them through the general county budget. In effect, this means that by paying into the general budget, cities with independent departments subsidize contract cities, which receive county services at less than their "real" costs.

Most arrangements do not have opportunities for subsidization. What occurs in the general case is best represented by Dan Grant's observations about Nashville Metro.

> In the opinion of this writer metro has already eliminated many duplications and has some economies to show for this, but, paradoxically, has caused increased expenditures which more than wipe out the economies.... It is therefore possible to contend logically that metro has eliminated the wasteful duplication of street maintenance and traffic engineer-

ing equipment in city and county departments (as it has), while also reporting that metro has caused new expenditures for a great expansion of
the street lighting program to light all highways out to the county boundaries. This is even more true in the case of school consolidation (1965, p.
40).

Juxtaposed against any savings from scale economies (see Proposition 1) is
the widespread evidence that reorganization significantly increases spending
for services. Examining Nashville Metro, Grant finds " ... bigger budget re-
quests and appropriations ... " (Grant, 1965, p. 40). Looking at educational
expenditures in Metro Toronto, Gail Cook concludes that the net effect of
federation was to increase the allocation of resources to education via an
expenditure increase (Cook, 1968). In his major study of Toronto Metro,
Kaplan (1967, p. 114) found that " ... the transfer of a service program
[from municipalities to the federated government] ... was usually followed
by an upgrading of service standards in some municipalities and an overall
increase in area spending for that service."

More in the realm of conjecture is the *cause* of increases in service spending. Grant argues that it is attributable to rising citizen expectations (Grant,
1965, p. 40). Kaplan (1967, p. 114) suggests that the reason is the expectations of the new, professionalized bureaucracy, which has powerful interests
in guaranteeing a secure political base. A possible third reason, founded upon
the frequent observation that service levels in the subunits do not fall below
their pre-reorganization levels, is that for a host of reasons decision makers act as
if they were Pareto-optimizing: increasing the satisfaction of some without
decreasing the satisfaction of any. The prereform levels become, in effect, the
floor for postreform service provisions. When this goal is coupled with the
forces encouraging service equalization throughout an entire area, or defin-
able zones—as in the case of differential taxing and service zones—the end
result is the following: the service floor for the area or zone is governed by
that of the subunit with the highest previous level. This helps explain why
service volume, and associated costs, increase. The existence of differential
taxing and service zones aids us in explaining why service inequities seem to
persist (see Propositions 13 and 14). Additionally, however, although a higher
minimum service level is defined by the Pareto-optimizing procedure, there is
no specified maximum level. Patterns of inequality may remain largely unchanged, while the entire distribution merely moves upward to a higher position, the permissible floor of which is defined by the subunit with the highest
previous level.

If economies of scale are realized when functions are transferred to an
areawide unit, this advantage should be contrasted with the evidence of significant increases in service spending. When the average citizen examines his tax
bill, he has little evidence that economies are taking place. This may partly

explain why citizens do not perceive the greater efficiency and economy (McArthur, 1967) that knowledgeable officials perceive (Grant, 1969).

Proposition 11: The creation of reformed metropolitan institutions increases the financing capabilities of governments in the area if the areawide institution is granted some fiscal powers, although citizens are no more willing to tax themselves in the larger entity.

The clearest example is in Toronto. It is said that when Metro assumed outstanding bonded-indebtedness for municipalities and took over the issuance of bonds, the savings reached about $50,000,000 through 1962 (Smallwood, 1963, pp. 12-13). An analogous situation exists in regard to the role of COG's in obtaining federal financing. Even before the requirement that the COG's approve projects of unit governments, the Supervisors Inter-County Committee in the Detroit area enabled participants to obtain more funds for public works projects, and in the Washington, D.C., area the U.S. Public Health Service provides extra funds and staff to the Council (Thomas, 1967, Table 92; p. 612).

Proposition 12: Despite the normative push for parity among jurisdictions, reformed metropolitan institutions often do not provide services any more equitably than unreformed institutions in the same area.

In one metropolis, Toronto, differentiation in services provided within municipalities included in the federation actually increased after reform (Smallwood, 1963, pp. 17-28), because local performance varied more than had been anticipated. Metro's share of service costs did not keep pace with the burden carried by the municipalities themselves and there was no equalization of the assessment base of the 13 local units. In other metropolitan institutions the mechanism by which service inequities persist after reform is service zones. Baton Rouge, Nashville, and Jacksonville all established such zones in which levels of services and taxes are higher than for the total area. All the city-county consolidations have similar zones, in part because all include areas of low population density.

Within zones the new institutions attempt to provide services uniformly. In Nashville, the attempt to extend services into previously unincorporated portions of the zone temporarily reduced service levels for the old city of Nashville. Not only the level but the manner in which services are provided can be homogenized within zones. A loaded example appears in the operation of the Contract Services Plan, or Lakewood Plan, in Los Angeles County. There is tension between the sheriff's deputies, provided under the plan and the large minority populations of cities that buy the police services. Deputies are said to act on the basis of "Anglo" or "area" norms and not those of the local community which they police. The city of Irwindale, California, largely

Mexican-American, ceased to contract for police services and established their own police department at greater cost, because residents felt contract deputies looked down on Mexican-Americans. Similarly, the city of Compton, California, which is three-quarters black, is not expected to contract for this service (Ries and Kirlin, 1970). However, the very existence of zones leads to continued interzone service differences.

Proposition 13: Reformed metropolitan institutions do not redistribute wealth among elements of their populations.

Redistribution of wealth is both more possible in larger governmental units and more likely to occur at levels of government above the municipality (Fry and Winters, 1970, pp. 508–22). But, although the federal and state governments perform redistributive functions, little redistribution seems to occur *within* reformed institutions. At first glance, Los Angeles County's Contract Services Plan would appear to be an exception, given that there was redistribution from independent cities to contracting cities. This has resulted in a redistribution toward lower-middle-class whites (Ries and Kirlin, 1970). Yet, redistribution is only possible because of the voluntary nature of the Contract Services Plan. There are cities that do not contract for county services, preferring to provide their own. Where arrangements include all subunits within a larger area (such as a county), the proposition appears to hold: there is a general lack of redistribution, which is largely attributable to the emphasis on tangible, private goods and the persistence of differential taxing and service zones.

Impact Effects

Although process and output effects have been objectives of metropolitan reformers throughout the 1950's and 1960's, albeit with changes in emphasis, concern with the impact of metropolitan institutions upon citizen attitudes and beliefs reflects a new set of values only now taking form. Despite the fact that we have not fully met earlier reform objectives, a new set of values by which to judge institutions is emerging. Banfield describes the general value which seems to be appearing.

> If some real disaster impends in the city it is not because parking spaces are hard to find, because architecture is bad, because department store sales are declining, or even because taxes are rising. If there is a genuine crisis, it has to do with the essential welfare of individuals . . . not merely with comfort, convenience, amenity, and business advantage, important as these are . . . whatever may cause people to die before their time, to suffer serious impairment of their health or of their powers, to waste their lives, to be deeply unhappy or happy in a way that is less than human affects their *essential welfare* (1968, p. 10).

If the metropolitan governance process does in fact exist in part to provide for the "essential welfare" of human beings today, by what yardsticks is individual well-being to be measured? Measures of well-being might include both evaluations of social conditions and subjective satisfaction or dissatisfaction with the state of metropolitan life and government. A separate study would be required to assess the quality of life in cities before and after the development of metropolitan institutions. No studies of which we are aware focus on this aspect of performance for numbers of metropolitan units. Data by which to measure such conditions are not easily assembled or evaluated (Krieger, 1969). Similarly we know virtually nothing about the impact of institutional reform on citizen attitudes and feelings. We next suggest five propositions about the psychological side of such impacts, but they are so tenuous, perhaps even contrived, that we might better have stated only that we know nothing about the impact of metropolitan reform on people.

Proposition 14: The objective performance of metropolitan institutions has little effect on citizen interest in or satisfaction with areawide institutions. Salience, interest, and satisfaction are a function of life space and life experience, not service levels.

In Nashville, a 1967 survey designed to elicit reactions to the 1962 reform indicated that the only variables significantly related to satisfaction with Metro were educational achievement and rural values. Satisfaction increased with increased education and decreased with the strength of rural attachment. The same general relationship occurs with regard to evaluation of services, taxes, and access to local officials (McArthur, 1967). Similar findings have recently been reported for the salience of American state politics (Jennings and Zeigler, 1970, p. 535). A related factor may be that the commitment to noncontroversial programs and professional expertise leads to an emphasis on upgrading performance through techniques largely inconspicuous to the public, such as data processing, improved communications networks, personnel, and budgeting policies.

Proposition 15: As with the case of performance, the nature of the decision-making process in the reformed metropolitan governance system has little impact on citizen interest or satisfaction with the new institutions and processes.

The literature on metropolitan reform does not deal directly with the effect of decision-making types on public attitudes. It appears, however, that regardless of how powerful or charismatic metro leaders are, their emphasis on unanimity and parity (Proposition 2) leads to consensual and noncontroversial public politics (Kaplan, 1967, p. 120). It is instructive that in state politics, people living in states with powerful governors are no more likely to follow state affairs or be interested in state politics than persons in states with weak chief executives (Jennings and Zeigler, 1970, p. 534).

Proposition 16: Reformed institutions are associated with lower citizen participation in metropolitan politics.

The evidence is sparse, but it all points in one direction. In Miami, voter participation has declined since the advent of reform, a decline which has not occurred in unreformed urban counties in Florida. Comparing average turnout in three prereform elections with average turnout in three postreform elections, a decline of 20 percent occurred in the first primary and of 14 percent in the second. Total voter registration has also declined (Glendening, 1967). In Toronto, neighborhood groups directed their activities to ward municipal governments, not Metro (Kaplan, 1967, p. 172). One study, based on an index of participation, concludes that neighborhood groups took a position on only 20 percent of the issues discussed (Kaplan, 1967). This conforms to the predictions of many commentators on metropolitan reform (Cloward and Piven, 1967; Banfield, 1968; Wood, 1959; Adrian, 1967). A suitable test of this proposition has not been made, however, because there are no studies providing analytical control concerning the appearance and disappearance of the crisis that generated reform in the first place.

Proposition 17: Reformed institutions reflect local political culture, but do not substantially change it.

The equally sparse evidence on this point indicates that the establishment of metropolitan areawide institutions does not lead residents to hold different views on the appropriate regimes or scopes of public policy. In Toronto, Metro operates in a way reflective of currents in extant culture which can be characterized as unitary and emphasizing a parliamentary mode of political decision making (Kaplan, 1967, pp. 70-80; 209-11). Local culture in Baton Rouge dictated that particular offices remain elective and others appointive (Harvard and Corty, 1964, p. 39). Similarly, in Dade County the "Miami milieu" which engendered reform remains unaltered by it (Sofen, 1963, pp. 74-91). The fact that metropolitan reforms vary so in form suggests the actual strength of the impact of local factors, including political culture.

Proposition 18: After the establishment of reformed institutions of metropolitan government, the primary local political attachment of citizens remains submetropolitan.

The question of under what conditions a territory legally defined as a unit comes to constitute a community with a sense of collective self-identity is extremely complex. Nevertheless, the relative salience of political jurisdictions to the metropolitan citizenry is very important in assessing their support for the activities of various institutions in the area. Attachment of citizens to submetropolitan units is one of the primary obstacles to creating areawide institutions (Bollens and Schmandt, 1970, pp. 298, 429), and the creation of an areawide unit seems not to cause a transference of attachment (Kaplan, 1967, pp. 218-20; Aron, 1969, pp. 57-81; Thomas, 1967, pp. 190, 677). Knowledgeable officials interviewed in four metropolitan areas indicated that local ties do not decrease, because cities are already too big for this kind of pride; that is, there was no community spirit to begin with (Grant, 1969). It is possible, however, that local identification increases, because metro makes real the threat to local community and the need for countervailing power.

Implications of Postwar Reforms: Can Something Be Done?

An optimal conclusion would assess the probable benefits and costs associated with alternative institutional arrangements and goals as a basis for policy. Then we might be able to decide, for example, under what conditions the net benefits obtained from the use of metropolitan special districts outweighed those obtained through choice of a federal arrangement, and for what narrower purposes one might be superior to the other. Certainly one cannot expect mechanically applicable algorithms with which to make social and political decisions. Ordinal (more or less) assessments of cost and benefit probabilities are the best that probably can be expected. Nevertheless, the utility of even a primitive balance sheet is apparent.

The evidence used here to build propositions on the effects of metropolitan institutional reforms is not sufficiently precise or extensive to assign probable costs and benefits accurately to alternative metropolitan institutions. Moreover, we are limited by the difficulty of creating comparisons between the performance of metropolitan and nonmetropolitan arrangements or of pre- and postreform situations. Given these limitations, a summary perusal of the evidence on institutions and effects suggests three general conclusions. First, institutional reforms in metropolitan jurisdictions have been evaluated through a shifting set of goals. It appears that in a given era, or perhaps in a given social group, one set of concerns assumes preeminence, to the partial exclusion of others. The 1970's will probably be keyed to impact goals, but we do not predict what will happen next. New values could appear, or the present set may recycle. Clearly, however, political institutions are unlikely to stand or fall over time on the basis of a single goal.

Second, the overlapping and recycling of goals suggests that the critical questions have to do with trade-offs when institutions are designed. In assessing metropolitan institutions, we reach the following conclusions:

1. The increasing impact of professionals upon policy making has been a nearly universal effect of metropolitan reform.

2. In general, few scale economies have been associated with reform, except for the metropolitan special district.

3. In the short run, the access of minorities is guaranteed, but may in the long run be diluted.

4. The roles and norms of official decision makers do not guarantee any substantive conception of the areawide public interest. All that is guaranteed is a more minimal, procedural definition governing public policy making.

5. Performance levels tend to rise, but so do the fiscal burdens accompanying increased expenditures.

6. The functional emphasis upon tangible goods, rather than upon amelioration of social problems, remains unchanged.

7. There is no immediate, short-term impact upon the distribution of power and wealth.

8. By and large, restructuring is associated with reduced citizen participation in the local electoral process.

9. The character of citizens' understanding of and attitudes toward the local and metropolitan political process remains largely unchanged, being more a function of life styles and experiences.

Inevitably, we come back to the question of whether institutional manipulation is a fruitful strategy for doing something about public problems in metropolitan areas. Our analyses do not provide us with a definitive answer. Adequate measures of the performance of metropolitan institutions do not exist, in part because study has focused more on promoting reform rather than assessing it. Furthermore, there are different interpretations for the observed effects of metropolitan institutional reforms. Each can lead to a slightly different conclusion.

To us it appears that the improvements in performance, great as they have been in particular spheres, have not reached expectations, in part because these expectations have changed in character over time. In the spheres in which institutional manipulation has had least impact, altering institutions means dealing with the wrong end of the causal chain. The institutions and patterns of behavior associated with them are consequences of more basic attitudes and social patterns. The value system or the distribution of power and wealth in large part determines the impact of institutional manipulation: therefore, these basic elements remain unchanged by minimally changing structures.

Third, it is apparent that in many metropolitan reforms the environment so predetermines the feasible limits of institutional alteration that metropolitan reforms really change very little. It should be emphasized that we are, after all, dealing here not with major structural changes but often with only the mildest reforms. Relationships within many reorganized systems are remarkably like those in the preorganization world. Metropolitan institutional reform has changed little, not because it cannot but because little has in fact been attempted.

BIBLIOGRAPHY

Adrian, Charles. "Public Attitudes and Metropolitan Decision Making." In Thomas Dye and Brett Hawkins (eds.), *Politics in the Metropolis.* Columbus, Ohio: Merrill, 1967, pp. 454–71.

Advisory Commission on Intergovernmental Relations. *Factors Affecting Voter Reactions to Governmental Reorganization in Metropolitan Areas.* Washington, D.C.: ACIR, May 1962.

Aron, Joan B. *The Quest for Regional Cooperation: A Study of the New York Metropolitan Regional Council.* Berkeley and Los Angeles: University of California Press, 1969.

Banfield, Edward C. "The Politics of Metropolitan Reorganization." *Midwest Journal of Political Science*, May 1957, pp. 77–91.

———. *The Unheavenly City: The Nature and Future of Our Urban Crisis.* Boston: Little, Brown, 1968.

Benson, Charles S., and Lund, Peter B. *Neighborhood Distribution of Local Public Services.* Berkeley: Institute of Governmental Studies, 1969.

Bollens, John C. (ed.). *Exploring the Metropolitan Community.* Berkeley and Los Angeles: University of California Press, 1964.

———. "Metropolitan Special Districts." *The Municipal Yearbook 1956.* Chicago: International City Managers' Association, 1956, pp. 47–55.

———, and Schmandt, Henry J. *The Metropolis: Its People, Politics and Economic Life.* 2nd ed. New York: Harper and Row, 1970.

Burton, Richard P. "The Metropolitan State: A Prescription for the Urban Crisis and the Preservation of Polycentrism in Urban Society." Washington, D.C.: Urban Institute Reprint 59-116-51, 1971.

Cloward, Richard, and Piven, Francis F. "Black Control of Cities." *New Republic* 157 (September 30, October 7, 1967): 19–21, 15–19.

Committee for Economic Development. *Reshaping Government in Metropolitan Areas.* New York: CED, February 1970.

Cook, Gail. "Effect of Federation on Education Expenditures in Metropolitan Toronto." Ph. D. dissertation, University of Michigan, 1968.

Dahl, Robert. "The City in the Future of Democracy." *American Political Science Review* 61 (December 1967): 953–70.

Douglas, Peter. *The Southern California Association of Governments: A Response to Federal Concern for Metropolitan Areas.* Los Angeles: Institute of Government and Public Affairs, University of California, Los Angeles, May 1968.

Dye, Thomas; Liebman, Charles; Williams, Oliver; and Herman, Harold. "Differentiation and Cooperation in a Metropolitan Area." *Midwest Journal of Political Science* 7 (May 1963): 145–55.

Erie, Steven. "The Los Angeles City Council: Some Determinants of the Locally Oriented Councilman." Senior Honors Thesis, University of California, Los Angeles, May 1967.

_____. "The Southern California Association of Governments: A Search for Effective Regional Leadership." Unpublished manuscript, University of California, Los Angeles, August 1968.

Fry, Brian R., and Winters, Richard F. "The Politics of Redistribution." *American Political Science Review* 64 (June 1970): 508–22.

Glendening, Parris N. "The Metropolitan Dade County Government: An Examination of Reform." Ph. D. dissertation, Florida State University, 1967.

Grant, Daniel R. "A Comparison of Predictions and Experience with Nashville 'Metro'. " *Urban Affairs Quarterly* 1 (September 1965): 35–53.

_____. "Political Access under Metropolitan Government: A Comparative Study of Perceptions by Knowledgeables." In Robert T. Daland (ed.), *Comparative Urban Research: The Administration and Politics of Cities.* Beverly Hills: Sage Publications, 1969, pp. 249–71.

Greer, Scott. *Metropolitics.* New York: Wiley, 1963.

Hanson, Royce. *Metropolitan Councils of Government.* Washington, D.C.: Advisory Commission on Intergovernmental Relations, August 1966.

Harvard, William C., and Corty, Floyd. *Rural-Urban Consolidation: The Merger of Governments in the Baton Rouge Area.* Baton Rouge: Louisiana State University Press, 1964.

Hawkins, Brett W. *Nashville Metro: The Politics of City-County Consolidation.* Nashville: Vanderbilt University Press, 1966.

Hawley, A. H., and Zimmer, B. G. *The Metropolitan Community.* Beverly Hills: Sage Publications, 1970.

Hirsch, Werner Z. *About the Supply of Urban Public Services.* Los Angeles: Institute of Government and Public Affairs, University of California, Los Angeles, 1967.

Jacob, Herbert, and Lipsky, Michael. "Outputs, Structure, and Power: An Assessment of Changes in the Study of State and Local Politics." *Journal of Politics* 30 (May 1968): 510–38.

Jacob, Philip E., and Teune, Henry. "The Integrative Process: Guidelines for Analysis of the Bases of Political Community." In Philip E. Jacob (ed.), *The Integration of Political Communities.* Philadelphia: Lippincott, 1964, pp. 1–44.

Jamieson, James B. *Park-Bond Voting in Municipal Elections.* Los Angeles: Institute of Government and Public Affairs, University of California, Los Angeles, March 1965.

Jennings, M. Kent, and Zeigler, Harmon. "The Salience of American State Politics." *American Political Science Review* 64 (June 1970): 523–35.

Kaplan, Harold. *Urban Political Systems: A Functional Analysis of Metro Toronto.* New York: Columbia University Press, 1967.

Kaufman, Herbert. "Administrative Decentralization and Political Power." *Public Administration Review* 29 (January 1969): 3–14.

_____. *Politics and Policies in State and Local Governments.* Englewood Cliffs, N.J.: Prentice-Hall, 1963, p. 28.

Krieger, Martin. *Social Indicators for the Quality of Individual Life.* Berkeley: Institute of Urban and Regional Development, University of California, Berkeley, October 1969.

Lansing, John B., and Marans, Robert W. "Evaluation of Neighborhood Quality." *AIP Journal,* May 1969, pp. 195–99.

McArthur, Robert E. "The Impact of Metropolitan Government on the Rural-Urban Fringe: The Nashville-Davidson County Experience." Ph.D. dissertation, Vanderbilt University, 1967.

Marando, Vincent. "Inter-Local Cooperation in a Metropolitan Area: Detroit," *Urban Affairs Quarterly* 4 (December 1968): 185–200.

Office of Executive Management, Bureau of the Budget. *Section 204 of the Demonstration Cities and Metropolitan Development Act of 1966: Two Years of Experience,* April 10, 1970.

Olson, Mancur, Jr. *The Logic of Collective Action: Public Goods and the Theory of Groups.* New York: Schocken Books, 1968.

Planning Commission of Nashville-Davidson County. *Local Government Fact Book.* Nashville: Central Printing Office, February 1968.

Press, Charles. "Concepts of Access and Representation in Metropolitan Area Studies." Paper delivered at the 1961 American Political Science Association meeting.

Rendon, Armando. "Metropolitism: A Minority Report." *Civil Rights Digest* 2 (Winter 1969).

Ries, John C., and Kirlin, John J. "Government in the Los Angeles Area: The Issue of Centralization and Decentralization." Paper presented to the symposium "Los Angeles: Metropolis of the Future?" jointly sponsored by the Institute of Government and Public Affairs and University Extension, University of California, Los Angeles, April 2, 1970.

Shapiro, Harvey. "Economies of Scale and Local Government Finance." *Land Economics* 39 (May 1963): 175–81.

Sharkansky, Ira. "Environment, Policy, Output, and Impact: Problems of Theory and Method in the Analysis of Public Policy." *Policy Analysis in Political Science.* Chicago: Markham, 1970, pp. 61–80.

Shoup, Donald C., and Rosett, Arthur. *Fiscal Exploitation of Central Cities by Overlapping Governments.* Los Angeles: Institute of Government and Public Affairs, University of California, Los Angeles, December 1969.

Smallwood, Frank. *Metropolitan Toronto: A Decade Later.* Toronto: Bureau of Municipal Research, 1963.

Sofen, Edward. *The Miami Metropolitan Experiment.* Bloomington: Indiana University Press, 1963.

Thomas, Nicholas. "The Roundtables of Metroplex: A Comparison of the Supervisors Inter-County Committee (Detroit), The Metropolitan Regional Council (New York), and the Metropolitan Washington Councils of Governments." Ph. D. dissertation, Syracuse University, 1967.

Williams, Oliver P. "Life-Style Values and Political Decentralization in Metropolitan Areas." In Terry N. Clark (ed.), *Community Structure and Decision-Making: Comparative Analyses.* San Francisco: Chandler, 1968, pp. 427– 40.

Wood, Robert C. *Metropolis Against Itself.* CED Supplementary Paper. New York: Committee for Economic Development, March 1959.

2 Metropolitanism and Decentralization

LANCE LIEBMAN*

Decentralization

The 1960's were an age of participation and decentralization, but these rhetorical doctrines did not flourish in a single form. In America the early decade saw a movement for participation because of its psychological values: alienated men, especially poor blacks, could be turned into healthy middle-class specimens only by being given control over their own destinies, especially control over formulation and management of the government programs aimed at helping them. More recently, decentralization has had a different focus: the growth and failures of national government—events especially marked in the very programs that sought to involve citizens—are asserted as reasons why government decisions should be made, and government programs managed by, smaller units.

The social propositions justifying greater political participation and decentralization are not peculiar to the American spirit or system. During the past 10 years there were similar movements in many countries, especially in Europe. Such political sentiments have been a basis for major government programs in Yugoslavia, England, France, Czechoslovakia, and Canada. In Europe *and* North America, participation and decentralization have been regarded as useful vote-getting slogans in local and national elections. Yet on the surface this localist urge—for neighborhood decisions, wide popular participation, and especially for the breakup of large municipalities—seems to run directly counter to the concept of metropolitanism: the central theme of the metropolitan movement must be that municipalities are too small and

*Assistant Professor of Law, Harvard Law School.

that larger units are required for effective governmental treatment of urban ills. This apparent incompatibility gives way, however, when one analyzes the confused and often contradictory statements of the decentralizers, and the conditions and concerns that have made their message popular.

Local Government as a Provider of Services

The chief activity of American local government is the delivery of services to its citizens. Counties, cities, towns, and villages extinguish fires, operate buses, collect garbage, and deter crimes. These functions, which residents could not easily perform for themselves in dense areas, gradually became public tasks during the nineteenth century. The quality of their performance today attracts most of the attention that Americans give their local government; and the existence of an urban crisis is usually blamed on alleged deficiencies in performance. Similarly, alleged declines in the quality of public service provide a principal support for arguments that cities should be broken up and their functions transferred to smaller units.

In fact, however, the diagnosis is crude and the prescription doubtful. Descriptions of urban life at earlier stages in American history abound, and all of the honest ones report that large areas of all cities were afflicted with terrible living conditions. There was no blissful period in American urban history, only periods when small areas of some cities were carefully tended and served, and when residents of those areas were not often disturbed by conditions elsewhere in the same metropolis.

In addition, an extraordinary combination of pressures has recently been placed on municipal services. First, urbanization has concentrated millions of poor people in the hearts of our major cities. Many have been black migrants largely unprepared for urban life, whose race insured constant friction with older residents. Inevitably, there is discrimination, which prevents adjustment and discourages effort.

Second, this particular period of urban growth has coincided with important changes in the national ethic: menial work is condemned from all sides; social and economic amenities only recently achieved by a fraction of the population now become sought by all; active protest, including violence, is applauded as an appropriate means of pursuing political and economic ends; patience vanishes. These changes are compounded by the saturation of society by television, which takes each "advance" in comfort and life style into every home.

Third, municipal labor relations have suddenly altered. Most public functions have traditionally been labor intensive. But until the 1960's, public employees worked long hours for low salaries. They may not have worked hard; their bosses may have been bad managers; but prodigality with labor could be tolerated because the wage costs were low. Suddenly, however, public sector wages in many cities are competitive with corresponding private employment. But work methods and management skill have not yet changed to take account of the increased cost of labor. Lateral entry, effective discipline, automation, and rational allocation of resources must be the response to higher wage levels in the long run. But some of the factors that brought about the rapid rise in municipal wages, such as the political influence of civil-servant unions, close relationships between supervisors and unions, and the short time horizons of politician-administrators, make managerial reform difficult and even unlikely.

In this situation, with the unit cost of public services rising rapidly while the demand for them is growing, the appeal of decentralization is understandable. The giant bureaucracies that provide services in big cities are cumbersome, woolly, and implacable. Nothing short of substantial restructuring seems likely to produce more service per public dollar spent. Decentralization is thought likely to result in increased efficiency for the following reasons:

1. The administrators of smaller units might get more work out of their men and machines. They might have a flexibility impossible in present city-wide departments. New methods of procedures could be tried in one area and prove applicable elsewhere. And it might be possible to measure and report publicly on the performance of local administrators, thus inducing them to increase their output.

2. Decentralization might in certain ways lead to a decline in the political power of municipal unions. Depending on the scheme pursued, unions might have less influence on the choice of local administrative heads, discipline (and reward) of workers might become easier, procedures for changing work rules might be simplified, and the power to shut down an entire city through strikes might be removed.

3. Decentralization would permit recognition of differential community preferences for municipal services: by permitting diversity of output, it could increase social efficiency by producing the services most desired by particular neighborhoods.

True, many of these gains might be achieved without breaking up the cities: significant progress toward flexibility and efficiency might be reached through administrative delegation of power and responsibility to carefully chosen local officials, while the perseverance and skill of elected and appointed officials could win back flexibility and discretion from the unions. But there is little evidence that this is happening. Cities of all sizes and histories, with all sorts of political traditions and with "reform" and regular mayors, seem to be losing ground, caught between increased demand and lower public productivity.

Then again, there is a real possibility that political decentralization could take place without bringing administrative gains. Citywide unions might be even stronger in confronting a variety of local employers; the lack of citywide services (accounting, consulting, planning) might be costly; an incompetent transition to the new scheme might replace a system that now more or less works with chaos. This last possibility is especially serious and cannot be considered unlikely. Very serious interests compete for urban benefits. Strong parties are involved, with much to gain from the present arrangements, with money to spend, knowledge of the system, and great patience. When similarly strong interests enter the fray in behalf of change—for example, groups with their roots in the civil rights movement or the poverty program—confrontation occurs. Confrontation is deplored as divisive, irrational, and destructive—and sometimes, as in the New York school decentralization wars, it is all of these. When new groups oppose what might be called the "Army of the Present," new schemes do not come into existence; rather they appear to be adopted but in fact turn out to be new names for old arrangements.

The realistic proponents of decentralization understand that they are proposing something far more than altered lines on organization charts. They know that public money is involved, as well as jobs and votes and honors, and that these are limited resources which can usually go to someone new only by being taken away from someone old. The proponents of decentralization are not saying that large cities cannot be governed or that smaller units more easily can be, but rather that the present system is deteriorating quickly and that the public pressure necessary to improve the situation cannot be focused on the kinds of details that would turn around the present municipalities. Instead, public pressure toward a broad and easily comprehended goal—decentralization—might allow radical restructuring, in the course of which more productive arrangements and relationships might be built. It is a gamble, they say, but one on which there is very little to lose.

Local Government as Decision Maker

City governments do not merely deliver services; they also decide public questions. Within constituted limits they decide what matters are appropriate for municipal decisions, what processes to use in making decisions, who should participate, and what the formal rules of the process should be. These public questions are often closely related to service delivery, blurring the distinction between the two municipal roles. For example, among the most important public questions are what services a city should perform, from whom the funds should be obtained for performing them, how services should be allocated among sections of the city, and who should be hired to perform the services. But municipal government also decides questions that are at best loosely related to service delivery: questions, for instance, of permitted land use, of income redistribution, of whether a thoroughfare should be closed for a patriotic parade, of what kind of private enterprises should be licensed. Generally, the decisions of a municipal corporation involve the allocation of some limited resource among its citizens, but the resource is often intangible and difficult to quantify.

The case for decentralized municipal decision making cannot rest on the ground that too few persons participate today in the processes by which cities make decisions. Power in virtually every city is widely dispersed. "Political machines" are in fact political sensors, responsive to individuals or groups who muster the energy to seek a particular point with determination, a little skill, and great patience. Decisions are made not by dictatorial individuals but by a process of bargaining which represents a great variety of interests in a continuing series of adjustments, compromises, and modifications. The argument for decentralizing this process must therefore include one or more of the following rationales:

1. Citywide bargaining structures are complicated. Although new groups can and do enter the process and exercise influence, many people are deterred from trying by the dense mysteries of the traditional arrangements. Decentralized institutions might—at least in the beginning—be easier to understand. Because the institutions would be geographically closer to the citizenry, more persons might seek to play a role.

2. The present citywide systems may not take sufficient account of local conditions, and may not be flexible enough to achieve different outcomes where varying circumstances require them.

3. The present systems overrepresent certain groups in the population and take too much account of their needs, conveniences, and desires as against the interests of other groups. Decentralized institutions are likely to be more responsive to groups that are too little regarded by the present arrangements.

This last argument is usually made by those who believe that blacks (or Chicanos or Puerto Ricans) are currently excluded from citywide power but could manage their own decentralized sections. More recently, however, the argument has sometimes been espoused by persons wishing to maintain some white-run areas in cities that have become heavily black. The argument is subject to question. Every study of political participation shows that individual participation is related to class, income, and family history. There is no evidence that significantly greater numbers would play an active role in a decentralized municipal system than are involved at present. Similarly, assertions that black (or white) influence would rise (or fall) are exceedingly doubtful. In one recent and much-publicized example, the decentralization of New York City's elementary schools, middle-class whites gained control of most of the local boards, including many boards in areas where they were a distinct minority. The same energies, ambitions, and commitments that lead to citywide influence produce local power. Also, as the election and representation processes of the Community Action and Model Cities programs have shown, the turnout for local elections is low, and locally chosen "community" representatives can be as cavalier toward their constituents, and as "unrepresentative," as "distant" City Hall functionaries.

The other arguments for a decentralized process of local decision making are stronger. Present processes are constipated. New departures can unbalance so many intricate relationships that it becomes virtually impossible to assembly enough of the relevant actors to obtain their agreement in the absence of a major (and unpublicized) crisis. On the vast range of important questions, no explicit consideration of policy choices ever occurs. When choices are made, they are often made by persons whose places in the bureaucratic structure assure constricted views of the relevant issues. An obscure example: those who negotiate wage agreements in New York City have never seriously sought productivity improvements. "Our job is tough enough as it is," they say. That leaves agency heads and their subordinates to try to increase output without being able to hold out a wage carrot. *Their* job, one might say, becomes not tough, but impossible.

Plainly, decentralization is no magic solution for the ills of metropolitan planning and law making. The central problem is that too many decisions must be made, and too many of these decisions are complicated and inter-related, overwhelming the capacity of any organization for an adequate weighing of choices and needs. Decentralization might, however, be useful in two ways:

1. It could transfer many decisions to a local level, at which level the decision makers (whether they are elected part-time representatives or appointed full-time officials and employees) could have a clearer view of the priorities of residents. No more decisions might be addressed adequately, but those that were might be the ones of greater consequence. No fuller list of the relevant factors might be considered, but those considered might be the ones of greater significance to the people who would be most affected.

2. A real shake-up of the system might produce, along with some degree of chaos, enough freedom from present cobwebs to make possible major improvement. (This is, of course, Jefferson's argument for a decennial revolution.)

A caveat is required. There are local leaders and local leaders. Today, in every city, persons with neighborhood roots and connections play large roles in citywide decisions. This characteristic of machine systems is equally an ingredient of anti-machine governments. The argument for a new scheme has to be either that new locals will play a large role, or that the old locals will play in different ways because the game has new rules. And it has to be argued that new people, if enabled to oust the incumbent local leaders, will behave differently and better. These arguments are plainly hard to support, but arguments asserting less are not worth very much.

Workers and Consumers

The strongest argument for decentralization may be one that combines both of the foregoing sections: municipal residents are unhappy about services and also about decisions (because they think they are related to inadequate services). The only approach with any chance at all of producing significant improvements, in a world that must continue to be short on public funds and long on public needs, is to increase the output per man hour of municipal employees. And the most likely (if not very likely) way to achieve that is to break up the big cities and in so doing to accomplish the following objectives:

1. To produce contact between employee and consumer, so that the ethic (the needs) of the consumer comes to influence the employee.
2. To increase the extent to which jobs go to the people, or at least the kinds of people, who live in an area so that the workers care about the level of service.
3. A transfer to a local level of decisions about which services are needed most so that social efficiency (satisfaction of strongly felt desires in relation to dollars spent) increases.
4. To place responsibility for service delivery on identifiable executives whose careers can depend on the results.
5. To place responsibility for policy questions, especially over service allocations, on political leaders whose careers can depend on public satisfaction with their decisions.
6. To shake up all processes—labor, budget, planning, politics—enough so that the newly responsible persons get some of the flexibility necessary to their success, and so that bottlenecks can be identified and public indignation mobilized against them.

Decentralization is probably necessary for these changes. Sadly, it is not sufficient. Its likelihood of achieving even significant progress in any single municipality cannot be high, because most people do not want to participate, only to complain; because many people cannot reach even the first level of sophisticated thinking about local problems (from "my street is dirty" to "my street is dirty because . . . and I will try to do something about it"); and because changes will take power away, or threaten to take it away, from groups and individuals whose present authority is both too small to let them think placidly about losing any of it and too recent for them to have forgotten what life is like without it. These groups will fight, and will have excellent resources.

But it is hard to see an alternative course of public action with even an equal chance of success.

Metropolitanism

Decentralization and metropolitanism are potential allies. The appeal of each is based both on rationalized service delivery and on a more appropriate alignment of the local political community. Believers in decentralization speak of more efficient management of municipal functions now miserably (but expensively) provided by ancient bureaucracies; believers in metropolitanism speak of replacing those same bureaucracies with new, areawide organizations that can draw on the achievements of twentieth-century management and organization. Decentralizers seek neighborhood institutions that will induce participation, provide flexibility, and bring local citizens into contact and trust with one another; metropolitanites not only see the more effective production of citywide public services, but fear the political separation of poor, big-city blacks from middle-class, suburban whites; and they hope that a spirit of community, and more enlightened multi-faction majorities, will follow an expansion of municipal boundaries.

Services

As to public service delivery, the compatibility of the bigger-city and smaller-city views is hardly labored at all. Public services are now provided badly, by the standards of not-so-efficient private enterprise and of large cities in Europe and Japan. The reasons are tradition, unions, and the absence of mechanisms for obtaining competent, middle-level managers and for providing the tools, discretion, and accountability that would let them get results. If we must assume that decisions are and will be made by political figures whose time horizons will never be long enough to let them be attracted to the necessary kinds of reform, and if we assume also that the citizenry, so vocal at demanding services, cannot focus its gaze effectively on the changes that might procure then, then structural reform—the most extensive reworking possible of the present patterns of public service delivery—seems vital to the production of high quality services at acceptable costs.

If General Motors or Gosplan were picking up urban garbage, they would assign the task to a regional division with authority to coordinate large resource decisions, plan certain activities (especially disposal), and reach labor agreements. They would also create subdivisions—perhaps in units of 50,000 to 100,000 population—with specific operational responsibilities, and then appraise, test, and compare their output. Most municipal services should be handled in this way: police, fire, sanitation, health, and education. For each service there is great need of appropriate local discretion and local citizen role; but it is hard to see any use, in achieving these local arrangements, for a unit corresponding to the present large municipality. Similarly, there is great use for a large unit, for planning, information circulation, and (with some services) control, but the present municipal boundaries are inappropriately small.

All of this is not to say that the withering away of the large city is imminent. Rather, were the nation establishing a fresh mechanism for the delivery of public services in today's metropolitan areas, it would probably choose a basic two-tier regional structure, with the metropolitan region subdivided into neighborhood districts. This arrangement has been established for Greater London, after the sort of consideration and review that the English do so splendidly.

The two practical problems of a two-tier service system (just as they are the problems with most governmental arrangements) are labor and money: who does the work and how to raise the money to pay their salaries. Neither subject can be treated exhaustively here, but the outlines of the likely conflicts can be foreseen. The issues about funds are differential capacities to pay, differential needs, and differential desires. Schemes will be more successful as they contribute to local concern and local involvement. Such a goal suggests the greatest possible delegation of decisions to the neighborhood level: let neighborhoods decide how much they want of each local service, so long as they can pay for it; let them choose weekly or daily garbage pickup, a mix between street sweeping and trash collection, and whether they prefer to be wakened by early collections or to have traffic disrupted by midday ones. There is no reason why these decisions need to be made uniformly for a city or a metropolitan area. On the other hand, current municipal service delivery includes some degree of implicit income redistribution from rich to poor; indeed, the hope for greater redistribution is a frequent argument for metropolitanization. As neighborhoods have decisions delegated to them, hidden redistribution becomes harder, and if they are asked to raise and spend their own funds it becomes impossible. The complicating factor is that some services involve varying *need*: fire fighting must be done where there are fires, education where there are children. Others involve varying *desires*: expensive language teaching, or clean streets. Presumably a satisfactory system would be one under which *necessary* services were always performed, with funds raised on some equitable (definitional problems are begged here) basis; while *desirable* services were available as neighborhoods chose to tax themselves for them, with, in addition, some acceptable amount of redistribution. This problem could be vastly reduced, at least theoretically, by an adequate minimum income system. If income transfer arrangements gave each family a fair amount of money to spend, we would be happier about relying on their own choice of whether to purchase private or public goods. The problem of creating adequate institutions which would let people express collectively their public goods preferences would still remain. And even with adequate income minima, society might want to decide on a level above that of the neighborhood not to allow a neighborhood majority to choose certain options: no police or no schools, for example.

Labor relations present different problems. The central one is who is to be hired. This question has fueled much recent urban discontent, and its essential paradox is not easily resolved. Jobs and promotions ought to be apportioned objectively; those most qualified should be chosen. But deprived minorities need jobs, ought to find them in the public sector, and may sometimes be better able, subjectively, to perform a task requiring sensitive relations with the consumers of a municipal service. It is hard to see how a two-tiered metropolitan/neighborhood system would allow this conflict to be resolved more easily than would present arrangements. Perhaps the metropolitan agency would command a wider job pool, including suburban jobs for which suburban workers are unavailable. Transportation costs, which keep urban blacks away from suburban jobs today, would still deter them, however, and if the price of getting labor was acceptance of black residents, it is clear how the suburbs would choose. A second possible gain would be that the wider job scheme would make it easier for white workers to transfer to more salubrious environments, opening up positions for black newcomers. But the basic problem will be with us long into all foreseeable futures.

Everything so far discussed has ignored problems of transition, which would of course be very great. One might visualize a trade-off: intermediate disruption while old methods are transformed into new ones, counterbalanced by the potential this disruption may represent for lighting dark corners, removing cobwebs, and straightening crooked paths. Even an entirely ethereal essay can mention one very practical reality: the certainty that much will turn on the competence of the individuals who head such a transition. The fruits of their labor will be with their city for a very long time.

Decisions, Participation, and Politics

Speculation about the impact of metropolitanism and decentralization on municipal service delivery is hazardous, but it is concrete and obvious compared to predictions about their effects on political processes, participation, and social life in general. Yet, after the fullest confession of inadequacy, the attempt may be worthwhile.

No single pattern can represent the present political structure of large American cities. The dominant issue, however, is the same everywhere: race. In most large cities, the number of blacks and what the Census calls "other nonwhites" is increasing while total population within municipal boundaries is remaining the same or decreasing. We do not yet have adequate interpretations of the 1970 Census, but those interpretations will surely show the movement of middle-class and working-class whites to parts of the metropolitan areas outside the central city, and also the movement of those whites from the East and the Midwest to the West and (a new event in American history) to the urban centers of the South. Blacks, Spanish-surname persons, and Orientals are moving into the center cities from the rural South and Southwest, and from abroad, but are making few inroads into suburbia.

The political results of this demography are well known. Black and other minority participation in politics is very low, but the potential minority vote gets larger and induces adjustment, seemingly before it achieves its portended strength. (Also, in many cities there remains a white group, usually small in number and large in money and level of participation, that will exert itself on the "black" side in local affairs.) Thus we now see the famous "black mayors" phenomenon and—less noticed but probably more important—the first signs of black aldermen, school board chairmen, and Democratic Party officials. While this process is going on, disintegration of the traditional ward-based political organizations is also occuring; as the perception of black strength precedes massive black voting, so awareness that the machine cannot "deliver" ("X's club couldn't deliver the mail," one hears, as if that were easy to do) becomes widely understood a decade after it has been conclusively demonstrated. One cannot yet say whether poverty (and model cities) program structures are replacing political parties; except in the most simplistic sense, they probably are not. We have not been told very much by political scientists and sociologists about the new suburban institutions and life style, but it is likely that new groupings and institutions are in the process of developing, suitable to the attitudes and behavior of the new residents.

In this context, the strident tone of some metropolitanism debates is understandable. Blacks (really *some* blacks) are on the verge of taking power in today's cities, and they see widened city boundaries as a way for whites to defeat their efforts at the moment of victory by importing new and unsympathetic voters. In fact, of course, the process would mean the restoration to the city's politics of people who had voluntarily removed themselves. Proponents argue that control over decaying central cities is not worth much, especially as bankruptcy impends. To which the blacks reply that the whites never said that when they themselves were in control, and anyway large amounts of federal cash may be around some (mythical?) corner.

This ping-pong game, unlike some others, is quite unenlightening. It is not, however, irrelevant. At least some black leaders must oppose regional structures, and many suburbanites must worry about what they see as enforced immersion in problems and due bills from which they thought they had migrated. For both groups, the combination of larger groupings with smaller ones may be an ameliorating possibility. If it is clear that neighborhood units with real power are to be created, the blacks will know they are not to be entirely disempowered: that they will retain jobs, political bases, and opportunities to pursue programs relevant to the special needs of their people. Suburbanites on their side will know that they can retain control of certain decisions vital to the character and amenities of their communities. These euphemisms hide real conflict, of course. Suburban "character and amenities" mean middle-class white towns; whereas among the hopes of metropolitanism must be the achievement of areawide land use and housing policies that make the dispersion of poor blacks more likely. Whatever decisions are left with local neighborhoods, some powers now belonging to the large city governments will go instead to a new institution built upon a white majority.

But the decrepitude of present schemes is a fact, and it may not be long before people will risk even some of what they have for the possibility of something better. If so, what might the new system look like?

The first question must be whether neighborhood units will be the building blocks of new areawide institutions. The disadvantages are the great number of them that there are likely to be and the conceivable constitutional problems over apportionment. The gains are in public identity, knowledge, and comprehension—adding up to an increased possibility of wide participation—and these gains ought to be sufficient to carry the day.

The second question relates to the trade-offs between part-time citizens and full-time professionals as managers. Suburbs now use more of the former because they can get them. Regional units will need a great deal of professional tending, but may therefore become distant, cold, and unresponsive. Part of the solution may lie in paid part-time or full-time officials of neighborhood units who also play formal roles in the regional administration. These people will focus too narrowly on the needs of their own constituencies, but without them no one may focus at all on that.

Third is the relationship that will be necessary with the larger political aggregates. Whether the region alone should deal with state and federal authority is a question hard to answer in the abstract. (It must also be noted that a substantial number of appropriate American metropolises include land in two states. Perhaps a paper so visionary, or so ingenuous as to suggest abolition of today's large cities should go the rest of the way and abolish the states as well. But this is another argument.) The vitality of the local units would be served if these were permitted some outside contacts, at least with private foundations.

What has here been written about the making of public decisions is greatly oversimplified. One can see the complexity of the problem and the choices by focusing on a single decisional subject, land use. Jeffrey Wood has written:

> If we wish to decentralize, but do not want to leave local districts ulti-
> mately free to impose costs on the wider community (or free to forbid
> uses beneficial to the wider community) we can restrict local land-use
> decisions to those instances in which both benefits *and* costs are borne
> primarily by the individual district making the decision. Alternatively, we
> can try to ensure that local district calculations of gain and loss from
> land-use decisions take into account the gains and losses attributable to
> those decisions but felt beyond the district boundaries or affecting those
> other than the district constituency.[1]

Wood then concludes that an appropriate land-use system for a two-tier metropolis would in the first instance have local bodies making decisions (zoning
map amendments and special purpose variances), with the matter then potentially available for review at the higher level. He suggests:

> Under this proposed disposition of "appeals" from local planning deci-
> sions, if the initial decision of an individual district brings opposition from
> the residents or the planning units of neighboring local districts (as might
> happen in land-use decisions along district boundaries or which affect
> facilities in one district used by residents of adjacent districts) central
> review and decision would be mandatory. If, on the other hand, the op-
> position was solely from local district residents or groups, central review
> could be made discretionary, with the understanding, implicit or explicit,
> that central reversal of the local decision depended upon conflict of the
> local decision with expressed city development goals.[2]

Adoption of a two-tier system necessarily requires arrangements of this sort
for almost the entire range of public functions and decisions. Each of them
requires difficult substantive and procedural judgments.

Ultimately, the answer to all questions about government is politics.
Rightly understood, democratic politics is a noisy, clumsy process by which
citizen interests are transformed into governmental outputs. Structures can be
important. They can be clogged, as by malapportionment or seniority, and so
prevent economical adjustments and resolutions. They can be lengthy, and so
delay necessary decisions. They can be dense, and so deter inexperienced

---

[1] Staff paper by Walter Fair, Jeffrey Wood, and Lance Liebman, "Decentralization of
New York City" (Association of the Bar of the City of New York, 1971).

[2] *Ibid.*

entrants and impose heavy transaction costs. Structures influence, and are influenced by, the political behavior that feeds into them.

The political price of present metropolitan structures is high. Blacks are entering particular games just as the benefits of those games are expiring. They are obtaining chips not redeemable in valuable coin. Meanwhile, the middle class is encouraged to do just what it would like to do, and what is most harmful for the system as a whole, to remove itself from the sections where the poor live, breaking off relations with those sections and their problems. In return, the middle class increasingly relates to poor neighborhoods—financially and politically—through Washington, a long-distance transaction that involves a minimum of flexibility, personal contact, and intergroup education. The promise of metropolitanism is the opportunity for local resolutions of local problems, for a new calculus of offers, acceptances, and contracts, under which—by what Professor Dahl sees as Madisonian democracy[3]—goals desired intensely are obtained at the price of lesser aims. Relationships of this sort could come about only clumsily and agonizingly, if at all, if one participant is the present central city, accustomed to dominate its surroundings, get most of the media attention, and function oblivious of its neighbors (and becoming accustomed to being regarded as incompetent, strike-prone, and broke). That kind of city is both too strong and too weak for useful relationships with a metropolitan area. The alternative is localism, whereby different areas within the old city are able to function by themselves in the regional amalgam, sometimes allying themselves with the other units of the former municipality, at other times reaching out for new alignments when part of the city and part of the suburbs seek common goals. One cannot say that such a structure would be more "liberal," "open-minded," "integrationist," or even "fair" than the present system. One can expect that it would be more flexible, more available to the efforts of individuals and groups not now active, and more capable of finding alternatives and possibilities excluded from the present parlance of political discourse. Finally, given the obvious dissatisfaction with the present arrangements, one can see very little reason not to shake things up and try again.

[3] Robert Dahl, *A Preface to Democratic Theory* (Chicago: University of Chicago Press, 1956).

3 A Federal Role
in Metropolitanism

CHARLES M. HAAR*

To some analysts of the current national scene, the urban crisis is fated to wither away; indeed to the degree that it can be said to exist, it is, in their view, but the creation of academics removed from reality, the gleam in the foundations' collective eye. To others, the unheavenly city is a product of professional reformers or bureaucrats who seek an outlet for their talents, or seek jobs, through a ceaseless and restless search for, and definition of, problems. And, most recently, the view of some urban watchers is that expectations keep rising and, presumably, must remain forever unsatisfied.

To those of the Dr. Johnson school, the large stone does exist. In analyzing grand concepts, be they the Holy Alliance, the Common Market, or federalism, an occasional descent to the field in which they operate and to the people upon whom they have impact can give the commonsense refutation to propositions like the nonexistence of matter. In discussing the legitimate and proper scope of federal activities in metropolitan governance, it seems particularly appropriate to assume that vantage point. For this purpose, a relatively recent New Jersey case is most helpful: *Vickers* v. *The Township of Gloucester*[1] is a prototype case. The challenge to the ordinance there at issue gave rise to reflections by the New Jersey court which set an apt framework for evaluating the federal response to metropolitan externalities, be they economic, social, or political. It also is helpful when discussing the extent of

---

*Professor of Law, Harvard Law School.

[1] 37 N.J. 232, 181 A. 2d 129 (1962), *cert. denied and appeal dismissed*, 371 U.S. 233 (1963).

federal entry into metropolitanism, in another and paradoxical sense: in a world in which all solutions are so hard to come by, it is at least a source of wry satisfaction that the problem is a genuine one warranting the anguish of thought and commitment.

A History of Prohibition

The Township Armed

In 1947 the Township of Gloucester, New Jersey, enacted an ordinance which prohibited the use of trailers for residential purposes within its limits.[2] The 23-square-mile township was located on Black Horse Pike, "one of the major traffic arteries in South Jersey" leading to Camden and Philadelphia, 10 to 18 miles to the southeast; yet at that time the land was largely vacant and undeveloped. The area claimed a population of about 7,000, up from 6,198 in 1940.

During the 1950's the urban fringe of the Philadelphia-Camden metropolitan region, constantly spilling over the surrounding political boundaries, extended out and reached Gloucester. Access to Philadelphia was improved by the construction of an overhead freeway connected with the Walt Whitman Bridge. Between 1950 and 1960 the population increased by as much as 121 percent, to a total of 17,500. The 10,000 new citizens, overwhelmingly white, native-born Americans,[3] settled, rather densely, in the northwestern half of the township, leaving the southeastern section largely untouched and vacant—nonetheless, this second section of the township lay clearly in the path for future development. Gloucester had not yet attracted any industry; thus the new people who came to establish their homes there had largely to commute to jobs in the nearby metropolis.[4]

[2] Trailers were not then what they were to become in the 1950's. "Often these units were without running water or sanitary facilities. There were no construction standards to insure even minimum protection against fire or collapse" (*id.*)

[3] In 1960, Gloucester township was 2.3 percent nonwhite; Camden city, only 5 miles to the northwest, was 23.8 percent nonwhite. The population of Camden decreased by 5.9 percent from 1950 to 1960 (statistics from the New Jersey Department of Conservation). Of the 568 municipalities in New Jersey in 1960, Gloucester township ranked 81st in population, but 345th in density (people per square mile). Gloucester township's average density of 749 people per square mile was not spread evenly throughout the township. Nearby Camden has a density of 13,467; New Jersey, with an average density of 808, ranked second in the nation behind Rhode Island.

[4] There were 6,060 employed persons living in Gloucester township in 1960. Of these, 2,488 were employed in manufacturing (the largest grouping, 419, worked in electrical machinery, equipment, and supplies); 447 worked in construction; 399 worked in retail trades, other than eating and drinking, and in food- and dairy-product stores; 480 were professional workers, most of them salaried; only 43 were employed in agriculture; and only 179 were unemployed (3 percent).

Between 1955 and 1957, 1,419 new homes were built. Mobile home vendors and mobile home park developers could be expected to fight for a share of the housing market, which was expanding fast as a result of the large influx of new residents. By 1957 their presence began to be felt.

In March 1957, Mrs. Laura Napierkowski purchased a mobile home for herself and her husband with the intention of placing it on her 4-acre lot. She was informed by Estella Brown, the township Clerk, that she could not do this; so she obtained a letter from her trailer vendor's attorney which stated, in effect, that the definition of "trailer" in the 1947 ordinance did not apply to the Napierkowski "mobile home."[5] However, the Township Committee, at its meeting on May 6, 1957, rejected this contention.

The Napierkowskis then obtained their own attorney, who entered into a series of unsuccessful negotiations with the Township Solicitor. On July 15, 1957, a suit was filed, alleging that the blanket prohibition of the 1947 trailer ordinance was unconstitutional. In response, the township declared, in part, that the proposed location of the Napierkowski trailer was a violation of the township's zoning ordinance.

The zoning ordinance, although a long-anticipated addition to the town code,[6] had actually been adopted only 15 days before Mrs. Napierkowski filed suit. It divided Gloucester Township into "A" Residence, "B" Residence, "C" Residence, "D" Residence, Business, Agricultural, and Industrial districts. The zoning map, dated September 1955, and last revised in May 1957, providentially located Mrs. Napierkowski's 4-acre lot in an "A" Residence district where dwellings were subjected to a bulk requirement that they consist of more than 800 square feet of usable first-floor area. The township claimed that trailers did not come within the definition of "dwelling," hence they could not locate in "A" Residence districts, and that even if they did in general come within the definition of dwelling, the Napierkowski trailer, which had a floor area of 500 square feet, did not meet the bulk requirement.

Nor did the township let matters rest there. Seeking to strengthen their case against Mrs. Napierkowski, the Township Committee, with the approval of the Planning Board, removed the complete prohibition on trailers. In its place "An Ordinance to Regulate and Control Trailer, Trailer Coaches, Camp

[5] Section 1 of the 1947 ordinance defined a trailer "to be a vehicle . . . designed to permit the occupancy thereof as a dwelling or sleeping place for one or more persons and having no foundation other than wheels, jacks or skirtings so arranged as to be integral with or portable by said trailer or camp car." Mrs. Napierkowski proposed to set her mobile home on a concrete foundation.

[6] Action culminating in the enactment of the zoning ordinance began with the creation of the Township Planning Board in 1953. Mr. Vincent Moffa was its first chairman. A continual member of the Board, he was serving another term as chairman in 1960 when the Vickers case was made a live issue in March and April of that year. Another charter member of the Planning Board was Mr. Robert Yost, Mayor of Gloucester in 1960, who served on the Board until 1956 or 1957.

Cars and Trailer Camps in the Township of Gloucester" was enacted on September 3, 1957. Henceforth, the parking, keeping, or maintenance of a trailer was to be permitted—but only in a licensed and regulated trailer park. Potential trailer park sites were confined by the ordinance to the "industrial" districts.

At this point the case went to trial. The court viewed the site and found not only that the Napierkowski neighborhood was rural but that it was likely to remain so for the foreseeable future. Accordingly, the court held, first, that the 1957 trailer ordinance was unconstitutional insofar as it attempted to regulate the immediate use of the land; and second, that the zoning ordinance was unconstitutional insofar as it regulated the use of a trailer on the Napierkowski property. The court held that the building code had no application to trailers and, in addition, declared that the 1947 trailer ordinance was unconstitutional in its entirety.

The township appealed. The Supreme Court of New Jersey certified the case on its own motion, and thereby brought the appeal directly to its own court.

The Supreme Court heard the case in March 1959, and on April 20, 1959, annulled the judgment of the trial court. However, it did accept some of the rulings of the lower court. For example, it upheld the trial court's determination that the building code had no application to trailers. It also sustained the finding that the Napierkowski property lay in an area which was rural and likely to remain so for many years to come. But the Court did rule that the application of the zoning ordinance to the Napierkowski property was a valid exercise of the zoning power: "Zoning must subserve the long-range needs of the future as well as the immediate needs of the present and the reasonably foreseeable future. It is, in short, an implementing tool of sound planning." The Court ruled, in addition, that the application of the 1957 trailer ordinance to the Napierkowskis' property was a valid exercise of the police power: "The use of trailers as permanent residences presents problems which are often times inimical to the general welfare."

The Court expressly did not rule on the constitutionality of the 1947 trailer ordinance and its complete ban on the residential use of trailers, since the ordinance of 1957 had repealed that earlier one.

Enter Harold Vickers

As soon as the decision was released, the Township Planning Board began contemplating in earnest the reestablishment of the "exclusionary wall."[7] This "rethinking" was perhaps spurred on because the township was now

---

[7]The court in *Vickers* uses this to justify in part its finding that the Board had not acted precipitously: "The exclusion of trailer camps had been a subject of frequent

beginning to receive applications for trailer camp permits. One such applicant was Harold Vickers, who had purchased 10 acres of "industrial" land in November 1957, about 2 months after the adoption of the 1957 trailer ordinance. The property, which cost him $10,000, was located on Erial Road (popularly known as Turkey Foot Road) and was 2½ miles from the closest home development then existing within the township.

Vickers moved his own trailer onto the property, and then submitted plans and application for a trailer court permit in September 1958.[8] In the fullness of time the Township Committee considered and rejected his application. In a letter dated December 8, 1960, the Committee explained that Vickers' trailer court would violate the township's subdivision ordinance, zoning ordinance, and building code.

Within a month Vickers brought suit against the Township. He alleged that he had complied with the 1957 trailer ordinance and he sought a judgment compelling the township to grant him a permit. In response, the township restated that Vickers would be violating the various ordinances and added the defense that his plans did not meet the requisite health standards.

The case went to trial on March 17, 1960, in the Superior Court for Camden County. The township now dropped its defense that Vickers would violate the subdivision and building code ordinances. But Vickers knew that he was at least partly in the wrong, and he asked for an opportunity to amend his plans so that they would meet the required standards of the State Department of Health. The Judge, W. Orvyl Schalick, granted the motion, and further, granted the township permission to "take such administrative action and review what action they may be desirous of taking in the interim time."

The Township Committee realized that Vickers would soon present it with plans which would comply with its 1957 trailer ordinance; the action it was "desirous of taking," therefore, was to repeal that piece of legislation and reestablish the complete prohibition which had existed before the maneuvers of 1957. There was another factor involved in this decision: 2 home developers had purchased a total of 500 acres immediately adjacent to Vickers' property, and the Committee was, unofficially, disposed to encourage them.[9]

---

discussion prior to the Napierkowski decision in 1959, and even more so since then . . . ."

[8] Relying on the township's defense, and the final decision, in the Napierkowski case handed down on April 20, 1959, Vickers acquired 10 more acres of wooded, industrial district land (across Erial Road from his first purchase) in May 1959. The deed is dated September 19, 1959.

[9] By the terms of the zoning ordinance, "A" Residence dwellings could be built in industrial districts but needed to receive permission from the Board of Adjustment. The question of rezoning the area from industrial to residential had not been put before the Committee even as late as June 1960.

The matter was discussed over the phone between the Mayor and Chairman of the Planning Board. The Township Solicitor was then instructed to draft the necessary ordinances so that they could be brought forward for Committee and Board action.

If we imagine a race, Vickers won. When the Township Committee convened its regular meeting on April 1, Vickers' amended plans were ready for presentation, but the ordinances were not. The Mayor fixed that. He referred Vickers' application to the appropriate town officials for consideration, then announced that the proposed zoning ordinance amendment "is being" submitted to the Planning Board for its considerations and comment at its next meeting, to be held on April 5, 1960. He proposed a motion to adjourn the Committee meeting until April 5, at which time the Committee would take up "any action regarding such proposed zoning ordinance amendment, and any other things that may be considered at the adjourned meeting." The motion was seconded and carried unanimously.

On the evening of April 5, 1960, the 2 major land-use control bodies of Gloucester Township held separate meetings in Municipal Hall. The Township Committee convened at 9 P.M. The record conflicts, but the Supreme Court of New Jersey believes that Estella Brown, the Township Clerk for the past 14 years, introduced for first reading "An Ordinance Amending an Ordinance Entitled 'Zoning Ordinance of the Township of Gloucester.' " Committeeman Calabrese moved, Committeeman McCann seconded, and it was unanimously carried that the ordinance be adopted on first reading and that the public hearing and second reading be held on April 22, 1960, at 8 P.M. in Municipal Hall. The Clerk then went on to introduce for first reading "An Ordinance to Regulate and Control Trailers, Trailer Coaches, Camp Cars and Trailer Cars in the Township of Gloucester." This time Committeeman McCann moved, Committeeman Calabrese seconded, and again it was unanimously carried that the ordinance be adopted on first reading and that the public hearing and second reading be held on April 22, 1960.

At 9:15 P.M. the Township Committee recessed. During this recess Committeeman Calabrese—also a member of the Township Planning Board—went to the Board meeting which had been in progress since 8 P.M. Again the record is not clear, but the Supreme Court of New Jersey states that he took along a copy of the zoning ordinance amendment and officially submitted it to the Board for "approval, disapproval, or comment."[10] Whether or not he complied with the ritual required by New Jersey State law, there is no question that the Planning Board was well acquainted with the matter to be placed before it.[11] The members had informally discussed the desirability of this amendment for over a year, and during the preceding 2 weeks Vincent

[10]Minutes of the Planning Board, April 5.
[11]Defendant's Brief.

Moffa, Chairman of the Planning Board, had talked over the proposed amendment on the phone with the Township Solicitor, who was at that point
drafting it. Mr. Moffa had also examined the matter with the Mayor and with
Karl Esler, the Secretary of the Planning Board. Indeed, only within the past
4 days he had unofficially received his own copy of the ordinance.

Mr. Calabrese remained at the Board meeting for only 45 minutes, returning to the Township Committee meeting when it reconvened at 10 P.M.
Although the Planning Board had not instructed him to do so, he orally
reported that the Board had unanimously approved the amendment. The
Committee then turned to other business.

The public hearing was held as scheduled on April 22. In response to
questions, Mayor Robert Yost, a Committee member for approximately 8
years, a member of the Planning Board from its inception in 1953 until 1956
(and, in his position as Mayor, a member ex officio for the past year and a
half)—"and who is a person to know such things"—declared that the "purpose
of the amendment was to protect property values, both present and future,
which might be adversely affected by a trailer camp." A letter from the
Planning Board, dated April 5th, was read. This letter restated that trailer
camps adversely affected property values; it continued:

> The Board believes that trailer camps do not contribute anything to the
> general appearance of the local scenery . . . [and] that the establishment
> of such camps would retard and, perhaps, choke the development of real
> estate for the area. To permit trailer camps . . . would not be in the inter-
> est of the general welfare of the community. The Board, therefore, regis-
> ters its approval of the amendment to the zoning ordinance amendment.

Upon second reading, both proposed ordinances were unanimously passed
and adopted.

The effect of these two ordinances in combination was to repeal legislation
adopted in 1957, which had permitted mobile homes within the confines of
closely regulated trailer parks in "industrial" districts (nearly the entire southern half of the township was zoned industrial), and to reestablish a complete
prohibition on the residential use of mobile homes, which had been township
policy from 1947 to 1957.

On June 20, 1960, an entirely new case was brought by Vickers to the
Superior Courtroom of Judge Schalick. From that point on, the case was
directed into the channels by which the judicial system reaches decisions,
channels which are sometimes unconnected with the substantive merits of a
case. Mr. Joseph Asbell, the plaintiff's attorney, cross-examined the Planning
Board's chairman in an attempt to draw out the story of how and why the
Board had reversed its policy on trailer camps. But the Court ruled:
" . . . whatever may have been the generation, the flowering into production
can only be by the action of the Planning Board minutes . . . not whatever

they may have discussed among themselves . . . the mere fact they discussed it has no probative value in the Court's opinion whatever, until the Board acts formally. Until then there is no action." Mr. Asbell did not subpoena the minutes of the Planning Board.

On Friday, July 1, 1960, Judge Schalick, ruled in favor of the township. Reasoning from the Napierkowski decision of 1957, he stated that the Superior Court

> . . . cannot separate some particular part of the community from the rest of that area and that we cannot particularly deal with just a limited area, but that there must be a standard adopted of treating the entire community, the entire municipality, particularly the main portion of the municipality where the subject premises are, in reference to the use of trailers and trailer sites. The Court finds this community is an expanding community, that this area may not be developing as fast as some others, but, nevertheless, there is a steady and possibly an advanced growth over the expectations we have had through the decade previously.

The Court further held that the Planning Board had had sufficient informal discussions on the zoning ordinance and that therefore the amendment could be taken as fully considered.

Vickers appealed. The Appellate Division did not reach the question of procedural regularity because they reversed the lower court's ruling on a point of substantive law. Judge Foley wrote for the Court:

> . . . it may be said that a ratio decidendi is whether the total prohibition bears a reasonable relationship to the purposes of zoning in light of the existing zoning pattern of the township, and the past, present and foreseeable future development of land use within its borders. To these factors must be added, among others, the area of the municipality, the size of its population, and in connection therewith the impact, if any, of regulated trailer parks upon land values and the general public welfare. . . . Viewing the facts comprehensively, we find it impossible to reconcile the complete exclusion of trailer parks with the rule laid down in cited cases. Surely, in this vast rural area, there must be some portion in which the operation of trailer parks would be compatible with the scheme of zoning the township has seen fit to select, and yet would not adversely affect existing or future uses of property located anywhere in the township, however zoned.

The township took its own appeal to the Supreme Court, and a final decision was reached on May 7, 1962. Justice Proctor, writing for the majority, ruled in favor of the township, specifically upholding the Superior Court's ruling. He stated that zoning must subserve needs even beyond the reasonably foreseeable future and that it is valid if it promotes the general welfare of the political unit which enacts it.

He emphasized that the court's role in reviewing legislative action on zoning ordinances was a strictly limited one. "If the amendment presented a debatable issue," he said, "we cannot nullify the township's decision that its welfare would be advanced by the action it took."[12] Then addressing his majority opinion specifically to the situation before the court, he concluded that if a zoning ordinance advanced a local community as a social, economic, and political unit, it was thereby furthering that general welfare requirement imposed by constitutional law. To the argument that total prohibition within the boundaries of a use was illegal, he replied: "We do not think that a municipality must open its borders to a use which it reasonably believes should be excluded as repugnant to its planning scheme."[13]

To these arguments, which to some ears may sound overly legalistic and doctrinaire,[14] Judge Hall issued an eloquent dissent. He was eager to look at the merits of the case, to cut through the procedural labyrinth, and to wrestle with the underlying issues.

First he stressed the land-use needs of a mobile, restless society. He pointed out the important place that trailers (or, rather, mobile homes—the euphemism that real estate people prefer) are occupying in the housing industry and in meeting the demand for moderate- and low-income housing. (We may note that they have increasingly become a way out from our current housing dilemma; in 1969, nearly 400,000 of the diminished total of 1,300,000 housing starts involved mobile homes.)

Judge Hall then went on to discuss his philosophy of planning and of metropolitan governance. He was troubled by the breadth of the majority's mandate in giving municipalities the freedom to erect exclusionary walls on their boundaries " . . . according to local whim or selfish desire."[15] Thus, the

---

[12] 181 A. 2d at 134.

[13] *Id.* at 138.

[14] Examples of arguments which may seem overly technical to the layman include the judicial presumption of the validity of legislative action which can be overcome only by an affirmative showing that it is unreasonable or arbitrary. Another apparent technicality is the provision of the New Jersey Constitution which provides for liberal construction of constitutional and statutory provisions concerning municipal corporations, in addition to broad powers granted municipal corporations which "include not only those granted in express terms but also those of necessary or fair implication, or incident to the powers expressly conferred, or essential thereto, and not inconsistent with or prohibited by this constitution or by law" (N.J. CONST. Art. IV, §7, par. 11). This is a more liberal attitude toward the power of municipalities than Dillon's Rule, formulated in *Merriam* v. *Moody's Executors*, 25 Iowa 163 (1868), which identified the municipal powers as (1) those granted in express words, (2) those implied in or incident to the express powers, and (3) those essential—not merely convenient—to the declared objects of the municipal corporation.

[15] 181 A.2d at 140.

prohibition of mobile home parks which symbolized a whole way of life and of governance, rather than the particular conditions of their actual exclusion, became his cause of concern.

As he ascended to the view from the metropolitan level, he determined that Gloucester Township was in the throes of an inevitable tide of decentralization, the mass migration of people from the densely settled centers of Camden and Philadelphia. Unlike the small, homogeneous communities already established around the central city which have an existing, and more or less permanent, character, such areas as Gloucester Township tend to be sprawling and nondescript, rural or semirural. It is their fate to be absorbed into the metropolitan melting pot. Their existing political boundaries, Judge Hall contended, are artificial; their land-use regulations express the sheerest localism, and their regulations constitute, in essence, a total disclaimer of the interests of the broader society. He conceded that such communities do not have to permit an Oklahoma land rush or a Western boom town; nor need they allow land to be used in whatever way strikes the current owner's fancy. But to control is not to exclude.

Therefore, Judge Hall concluded it is the judiciary's responsibility to supervise the zoning power to ensure that parochial and exclusionist attitudes do not run counter to the more general welfare, which must transcend the " . . . artificial limits of political subdivision and cannot embrace merely narrow local desires." The state judiciary is there precisely for that job: it cannot properly tolerate provisions designed to permit as new residents only certain kinds of people[16] or only those who can afford to live in certain types of preferred housing. Nor can it tolerate regulations aimed at keeping down the tax bills of current property owners.[17]

---

[16] There may be a constitutional question whether this complete restriction of mobile homes from the township infringes on the constitutional right to travel. In *Shapiro* v. *Thompson*, 394 U.S. 618 (1969), the Supreme Court held that the state welfare requirement of a residency of one year was a violation of the Equal Protection Clause inasmuch as the classification of otherwise eligible welfare recipients was an infringement on the right to travel interstate and did not support a compelling state interest. The argument could be made from this precedent that the complete exclusion of a mobile home park was a similar infringement on the Equal Protection Clause because the exclusion acted as a classification which inhibited interstate travel and did not promote a compelling state interest.

[17] The majority opinion indicated that local officials had said in private discussions that the people who lived in trailers were a shifting population without roots and did not make good citizens. The judge recognized that established residents frequently resent and distrust newcomers (especially apartment house dwellers) who vote on school budgets and for local officials, but who, they presume, do not pay sufficient real estate taxes to cover the costs of the local services they receive.

The Courts Arbitrate the Metropolitan Case

Since the *Vickers* decision, the suburban-central city split has worsened. The President's Task Force on Suburban Problems, reporting in 1968, emphasized that job opportunities (especially those in low-wage-paying manufacture) are increasingly leaving the central city. No longer is industry dependent on natural resources, on special ports, or on railroad marshalling yards; to the contrary, its land and transportation requirements are met more adequately in the suburban and exurban portions of the metropolitan area. Thus, the poor and the minority groups are being deprived not only of access to housing (if, as an earlier New Jersey case had put it, a suburb may be permitted to place a Chinese exclusionary wall around its borders) but of the very opportunity to make a living.

Judge Hall's approach to metropolitanism is especially noteworthy because he pulled out the issues in a way that earlier courts—and the rest of society for that matter—discreetly avoided. Wayne Township,[18] for example, in an earlier New Jersey decision, justified a minimum-dwelling-size regulation on the basis of health and safety, although one could argue convincingly that neither issue was the real motivation (the more forthright approach of the concurring opinion, which would have upheld the regulation on an aesthetic ground, was too blunt for its day). In state court opinions throughout the 1960's, the trend was to brush away the question whether the maintenance of property values was an appropriate exercise of the zoning power, to avoid any review of intended (or unintended but inevitable) consequences of regulations that excluded selected groups or classes of people, and also to shy away from evaluating the overspill effects of local regulations on neighboring communities and the states.

Both in its irresolution during the period of judicial scrutiny and in the way it finally came down, the Vickers decision is very much part of the mainstream of American decision making on zoning ordinances. As early as *Euclid* v. *Ambler Realty Company*, the very first Supreme Court opinion upholding the zoning concept, the court had stated:

> It is not meant by this, however, to exclude the possibility of cases where the general public interest would so far outweigh the interest of the municipality that the municipality would not be allowed to stand in the way.

But this flag, once hoisted by the Supreme Court, rallied few supporters in the judiciary. Despite growing sophistication about metropolitan networks,

---

[18]*Lionshead Lake, Inc.* v. *Wayne Tp.*, 10 N.J. 165, 89 A.2d 693 (1952), *appeal dismissed* 344 U.S. 919 (1953).

social cost-benefit accounting among local units of government, postures of presidential national commissions, exhortations by federal departments that suburbs should bear their "fair share" of the housing and welfare burdens, and the restlessness of minority groups and of central city mayors, few judges would grasp the thistle. Some *dicta* about the unreasonableness of large acreage zoning did appear. But, by and large, the quandary over the pattern of American settlements was left to other Alexanders.

As students of a realistic jurisprudence, then, we can only applaud Judge Hall's willingness to isolate underlying factors. But as one attempts to grapple with the Gloucester Township zoning dilemma, at least two puzzles emerge, and in the curious way that paradoxes operate, they blunt initial reactions.

First, who are the true parties to the conflict? For behind Vickers, a man striving earnestly to earn a profit by putting a trailer camp on his acreage, there march all the *amici*—the cities of Newark and Camden, which want to provide homes for their workers; federal officials, who believe that the housing, welfare, and education burdens are too heavy in the central cores; liberals, who wish to integrate suburban areas; mobile home manufacturers, seeking a mass market; and the poor—white and black—who find themselves locked into the central city. Quite an array, and seemingly all groups warrant protection by the judiciary. Yet, one must ask, how do these marchers get into the act? Or, as lawyers so picturesquely phrase it, where is their standing to sue? And, assuming their admittance, who speaks for them, asserts their interests, makes the reasoned elaboration of the case, presents testimony, provides the starting point for the judicial reasonings and conclusion, settles the terms of the final disposition? The courts, even though they provide the only local forum that does not depend upon the crisis constituency of political zones, cannot cope because they do not have before them all the parties who have a legitimate stake in the resolution. So, the difficulty with Judge Hall's dissent is that the court, under the present system, never gets a chance to deal with all the interests that should intelligently be taken into account.

Second, even though he might be alerted to the inter-community struggle, how ably could a law school graduate—even donning judicial robes—cope with it? Where could the judge learn of the existence and contents of a physical plan for the entire region (with ramifications extending from Camden to Philadelphia)? How would he learn whether the burden (and who would define it?) of low-cost housing was being cast unfairly on one unit of the metropolitan area, and how could he measure and assess such unfairness or disproportion? How could he perceive, and then go on to weigh, the reasonableness of a process of coordination among municipalities which was intended to provide suitable sites for houses to meet the needs of various levels of society, measured in terms of income, age, or size of family? If no published comprehensive plan existed for the entire metropolitan area, one that had been democratically participated in, voted on, and adopted, from where could he assemble one? Where, with only two parties before him, could a judge acquire the relevant information about the metropolitan trends in population, housing, transportation, and land development? Should one turn over such complex issues to a man with no training and no background for them and who picked up this case, rather than 50 divorce cases, because the Chief Judge of the local courts assigned it to his group?

There is still another troubling aspect: surely not every community need be a microcosm of the whole, lest we be ushered into that age of which de Tocqueville warned, an age of mass cities with no distinctive flavor, with interchangeable parts and characteristics. As a society are we to rule by some list, that every city shall have 15 percent poor, 30 percent lower-middle class, 45 percent upper-middle class, with the rest upper class—leaving scant room for variation, individuality, or personality, not even allowing for natural resources, topography, economics of scale, or factors of industrial location?

That goes too far, perhaps. Some of these dilemmas might be avoided if a court were simply to cut out the extremes of exclusionary local regulations, or to issue some form of generalized decree that cities by and large should be heterogeneous or, returning to Piers the Plowman of 1385, that the poor should be allowed to live near the rich. And halt there. For this rough fairness is attainable by the judicial system, and it could be the starting point for the democratic resolution of the patterns of population settlement, and for establishing the framework for the necessary logrolling among the metropolitan communities.

Remedy, then, is not without difficulties. In one sense, Judge Hall was saying that there had been an improper delegation of power, or, more precisely, a delegation without standards—in this case, from the state legislature to the locality. Thus if Judge Hall's reasoning had prevailed, the edict rather than fashioning a traditional decree would have taken the form of a criticism of the state zoning enabling act or of an appeal to the state legislature to rewrite its subdelegation of powers. But whatever the format, the metropolitan mismatch is not one that lends itself to easy judicial formulation. The notion of the interdependency of mankind does stir men's blood, but it gives only little guidance, unfortunately, toward defining what actions a court expects from Gloucester Township, New Jersey, vis-à-vis the rest of the interstate metropolitan area.

The Federal Government—Who Needs It?

The *Vickers* case merits such lengthy consideration because, apart from its intrinsic fascination, the dilemmas it raised are the very ones that we must resolve if we are to deal effectively with the urban crisis. They confront us all in dealing with the governance of metropolitan areas; and in this case, involving a zoning regulation, the issues of regional impact and of the external effects of local actions, are sharply posed, in a dramatic manner for analysis and, hopefully, for resolution. Can people residing within one locality make land-use decisions whose externalities bear directly on other people living in what is economically and socially the same unified territory, but one divided in terms of power into many separate territories by an imaginary line drawn on a map?

The debate of Judge Hall and his colleagues reminds us of the role of structure and of institution even in a society under the siege of technological and political change. Most of our citizens deem zoning ordinances—whose individual import seems so slight that their ramifications have been largely ignored by social scientists—the essence of grass-roots society, the birthright of the locality. *Vickers* shows how fragile is this comprehension. The cleavage between home rule and the increasing interdependence of our metropolitan society is poorly served by a federalism still rooted in 1789. The dilemma before the New Jersey Court calls for a land-use system which can provide an assessment of externalities not taken into account by the individual systems of each locality. Most difficult of all, at the base of such evaluation lies the need to develop some vision of the future of the total metropolitan area.

In short, the pivotal conflict on the American metropolitan scene is one between, on the one hand, such acknowledged verities as local rule, grass-roots knowledge, citizen participation, and community decentralization and, on the other, realization of the undesirable external costs imposed on others by local rule—both the wastages that ensue and the lost opportunities that might have been achieved through concentrated urban strategies. *Vickers* reminds us of another fundamental: unlike the philosopher, we cannot, having decided that equities are about evenly balanced, just silently steal away; the question demands resolutions; one side must win, the other lose.

In my judgment, a wise resolution to this conflict of values and institutional stalemate can come only through federal aid and federal action.

At first this proposition seems too stark. What can the federal government, with its lofty, sometimes disdainful Washington perspectives, its vast bureaucracies, and its own trifurcation of powers among the federal courts, the Congress, and the presidency, do about the 238 Standard Metropolitan Statistical Areas scattered across a vast continent? Nevertheless, I believe this is the appropriate response. And there are at least six considerations of the nature of the federal government which form the basis for my conclusion:

1. its interest in the efficient and economic distribution of monies;
2. its role as an institution builder and protector;
3. its role as an innovator in coping with metropolitan problems;

4. its role as an implementor of local and metropolitan policies;

5. its role as sponsor of national policies which require metropolitan cooperation;

6. its role as a clearinghouse and source of technical assistance.

Of course, what may currently be perceived as an institutional crisis may in fact be symptomatic of a far more primitive disease—an acute case of urban fiscal starvation. If this is indeed true, and the metropolitan issue is merely a crisis constituency of divergent forces, with no common center, then the federal government is missing the point when it seeks to revitalize the supply and delivery models, the current machinery of metropolitan governance. If, on the other hand, the crisis is not wholly a problem of resource shortages—and current appraisals of the governmental delivery systems evoked by the debate over revenue sharing do much to reinforce the conclusion that money alone will not save our cities—then the case for the federal government's pursuit of structural reform at the metropolitan level is significantly heightened. It is a belief that the second approach is the more correct one that underlies this listing.

One pervasive theme remains to be stressed: recent history and legislation show a steady, sometimes unexplored, at times divergent but nevertheless consistent trend toward creating and developing a federal role in metropolitan governance. Washington is apparently the governmental level most removed from metropolitanism; yet, the federal government—whether by design or by accident—has been the lifeblood of the movement. On the executive level, the Bureau of the Budget and the Department of Housing and Urban Development (HUD) have both encouraged areawide activities. Presidential messages have been sounding the theme of metropolitan cooperation. Congress, through its insistence on regional organization as a precondition for the availability of grants-in-aid, has given metropolitanism a boost that it would otherwise never have had. The attainment of enunciated national goals, whether these be desegregation or the establishment of public housing in previously barred portions of the metropolitan area, have been, at least partly, a federal responsibility and achievement. Indeed, the governmental institutions now most active in serving metropolitan needs, the nearly 200 councils of governments and regional councils, are products of federal initiative. And for this historical trend, I suggest, there are sound and seemingly irreversible reasons.

The Federal Government's Interest in the Efficient Allocation of Public Resources

One need hardly subscribe to Marxist political theory to begin an exploration of the federal interest in metropolitan governance by stressing the issue of money. With the recognition of the cities' starvation, with the emergence of revenue-sharing proposals—increasingly accepted (although in varying versions) by Republicans and Democrats alike—has come recognition of the federal government as the most ingenious and effective tax-collecting agency yet devised. It has been accepted that city problems cannot be struggled with or forestalled without massive federal aid. Before bloc grants appeared on the fiscal scene, a virtual deluge of federal categorical grants had already come forth to meet differing aspects of the urban dilemma. From some 40 grants at the start, the Great Society has multiplied its offerings to over 400, as different needs and different groups have vied for funds. Some $15 billions now go to state and local governments to aid them in the discharge of their services and functions.

At least two major consequences have ensued: the overlapping of funds and an occasional confusion and contradiction of program objectives. To bring some order into the grants and to coordinate their impacts, Congress in 1962 chose to employ one technique, the imposition of areawide requirements for the use of water and sewer, open space, mass transportation, and law enforcement grants; and it has continued to attach such requirements to various programs of metropolitan significance. Unless a comprehensive plan is devised on an areawide basis, therefore, no federal funds will be forthcoming for such public works. Section 702 of the 1965 Housing Act,[19] to take one example, states that a project cannot be funded unless it is consistent with criteria established by HUD for a unified or officially coordinated areawide water or sewer facility system that is part of the comprehensively planned development of the entire area. Once set, this has been the pattern of administration laid down in subsequent acts.

And, as the metropolitan criterion has been administered, many significant savings have accrued both to federal and to local governments, either because of lowered unit costs, because duplicate spending by federal and local government has been prevented, or because inconsistent and shortsighted developments have been forestalled. This need, therefore, to assume advance programming, economies through consolidation, and large-scale production and delivery of services to reduce unit costs, underlies the federal interest in metropolitan development. In a sense, the areawide requirement is Congress's way of rationing capital. With overwhelming demands come priorities.

But the federal legislator and administrator are now found with the very same dilemma that confronted Judge Hall, and it is one that lies at the heart of any job which reviews rather than initiates programs: how do you know that the areawide plan is more than a patchwork, that it is based on sufficient data, considers alternatives, is oriented toward problem solving, and is truly

[19]42 U.S.C. §3102 (Supp. V, 1970).

representative of the wishes of the community? How do you know that the framework of the metropolitan plan has been sufficiently articulated, so that it can act as a capital budget for the metropolitan area and can result in the wisest use of funds, insofar as areawide projects and priorities are concerned? What are the standards for review?

By an unexamined but natural accretion, the federal government seems to have become convinced over a period of 20 years that the conduct of the various branches of government in this country can be improved only by a process of rational forethought and analysis, commonly subsumed under the label "comprehensive planning." Hence, under Section 701 of the Housing Act of 1954 as amended,[20] vast sums of money (vast so far as the planning profession is concerned, minuscule by contrast to operations or expenditures of functional departments) have been expended on such planning. Congress has officially recognized the need for local communities, cities, counties, and special districts to plan together. If it is claimed that state planning and local planning should be aided by federal grants, it becomes even more appropriate that metropolitan planning should become a beneficiary of federal aid. Indeed, a sharper case can be made for the federal role here; in fostering an institution that would not otherwise be financed at all because it currently commands no representation—and without representation, taxation and money raising become impossible.

Under existing laws, the Secretary of HUD is authorized to make planning grants to governments and agencies to establish and to improve planning staffs and techniques on an areawide basis. In its mandate, Congress stated: "[The Secretary] shall encourage cooperation in preparing and carrying out plans among all interested municipalities, public subdivisions, public agencies, and other parties in order to achieve coordinated development of entire areas."[21] Order and coherence in the structuring and management of metropolitan areas as essential components of the federal system are thus the aims of the national government.

Comprehensive planning for urban development requires federal assistance to both state and local governments if it is to resolve the externalities that result from increased population concentrations in metropolitan areas. Within the existing local government structure, there is no means whereby one can even gather the data that would be necessary to evolve a comprehensive notion of the metropolitan problem, let alone a comprehensive plan for its amelioration or solution.

Alterations at the local level can come about only in the face of a crisis sufficient to mobilize the diverse constituencies; otherwise, we are con-

---

[20]40 U.S.C. §461 (Supp. V, 1970), *amending* 40 U.S.C. §461 (1958).
[21]*Id.*

demned (in an unassisted condition) to cyclical enactment of the crisis constituency syndrome. Local folks are too smart to pay dreamers for spinning solutions to problems that are only dimly perceived; but the federal government (either because it attracts a quota of dreamers or has better vision) will offer money for planning agencies and for the metropolitan plan. The plan can become the cause of the reformers at the stage when the crisis constituency is beginning to form. The plan can generate suggestions for improvement which were not born out of a crisis atmosphere. Thus, there is a commonality of goal from the outset; the plan is also a standard for measuring progress in reform.

With the encouragement of planning as an institution, there also come its inevitable counterparts—requirements. These might be regarded as federal intervention in an inherently local situation, but they might also fairly be understood as an assurance of a valid return for services rendered. They establish minimum planning performance standards. Thus, from the inception in 1954 of grants for comprehensive planning, there has been the requirement that a plan must include a comprehensive transportation component. Every local community that wishes to plan, so runs the legislative intent, has to develop long-range solutions for the movement of people and goods: without studying the needs and resources and the transportation goals of the metropolitan area as a whole, it will be difficult not only to plan wisely for the future of an area but even to plan adequately for a particular function—in this initial grant, the efficient use of monies on the interstate highway system and on the public transportation system is indispensable for the circulation of population in a metropolitan area. So if you take money to plan, transportation has to be included in your work schedule.

And again, as housing emerged as a primary national concern, the Housing and Urban Development Act of 1968[22]—that great, yet still latent, Magna Carta for urban development in this country—required that metropolitan planning include a housing element as part of comprehensive land-use plans. This requirement meant in effect that consideration was to be given to the projection of zoning, community facilities, and population growth trends, so that the housing needs of all classes of the population, at different prices and rentals, both of the region and of the individual local communities, should be adequately covered.

Constraints to low-income housing may vary from area to area: in some, the problem may be building codes; in others, it may be exclusive zoning ordinances; in still others, it may be a shortage of technological and management skill in rehabilitation; yet again, it may be labor union practices. Identi-

---

[22] 82 Stat. 476 (codified in scattered sections of 5, 12, 15, 18, 20, 31, 38, 40, 42 U.S.C.).

fication of the basic constraints and development of action programs become yet further aspects of institution building on the metropolitan level.

To insure that planning is not merely functional but involves an assessment and balancing of priorities—the issue for which there were admittedly no advocates in the New Jersey court—the federal government has stressed its comprehensive aspect. Problems (and solutions, hopefully) are to be viewed in a broad context and are to include an overview of the metropolitan area itself.

Once a commitment has been made to institution building, the federal government is called on to help decide who should be a proper recipient of these areawide planning funds. The legal status of the recipient agency, its in-house staff capability, its sources of local funding, its acceptability to the local units of government become proper matters of federal concern. Selection and representation processes are, of course, key issues. So, as the result of a 1965 amendment,[23] planning grants are authorized for organizations composed of elected public officials who are representative of the political jurisdictions within the metropolitan area. In many cases, councils of governments have been accepted by Congress, under present circumstances and constraints, as the most suitable instrumentalities. Federal money for studies, data collection, metropolitan plans, and program development has provided their financial mainstay. Now nearly 200 of these councils of governments are increasingly important in making decisions that affect people's lives and the uses of land in metropolitan areas.

At least one further development of the federal role in this area is worth noting. This relates to what makes any program go, the quality of the bureaucracies. The basic determinant of the success of a metropolitan agency— its capacity to stay relevant—hinges on the capability of its professional staff. Among the skills, experiences, qualifications, and budgets that might be employed, I wish to emphasize here only one factor that is of special federal concern in the training of future city and regional planners: the bringing of more members of minority groups into that profession.

A survey carried out by HUD during the last year of the Johnson administration indicated that only a little over 1 percent of all city planners came from minority groups. To meet this discrepancy, the then administrators of HUD created an urban management assistance administration; and programs were launched in cooperation with universities, public agencies, and professional societies. The program included work-study assignments with metropolitan councils of government and planning agencies at local and regional levels and also intensive university training. In the 1969–70 fiscal year alone,

---

[23] 40 U.S.C. §461(g) (Supp. III, 1968), *as amended*, 40 U.S.C. §461(g) (Supp. V, 1970).

approximately 150 graduate students, 200 undergraduate students, and 200 high school students took part in the program.

This is an area in which the federal government should be a decisive leader. One of the weak links in metropolitan planning and governments has been the lack of awareness of the attitudes and needs of the ghetto community. What is it that minority citizens expect from government institutions, which in the main they do not control? What manifestations of local governmental power are most feared, or regarded as most debilitating, by minority groups? I regard the introduction of minority group members into the planning process, and the feedback of their findings and perceptions, as perhaps the single most valuable improvement we can make today in sound metropolitan development. The involvement of men with a sense of cause, with a unique insight into the socioeconomic phenomena of the metropolitan region, can have a profound impact on the philosophy of metropolitan planning and on metropolitan action.

The federal government is thus providing the funds and encouragement to develop those comprehensive plans for the metropolitan area (and those metropolitan planning, coordinating, and budgetary agencies) which Judge Hall could have used as a referent in his dissenting opinion. An intelligent response from the public sector requires planning, a planning sufficiently comprehensive to coordinate all the available resources and to account for all the major components of the problem, so that a partial solution will not fall prey to the vicissitudes of unincluded factors and dimensions. And, of course, once this analysis, the weighing of alternatives, and final formulation have become part of local government activity, many other consequences for expenditures in metropolitan development will emerge.

In essence, the federal establishment, by virtue of being in, and yet apart from, any local situation, is in a position to help foster the institutional arrangements which can be the hallmarks of a viable system of metropolitan governance and which can provide a response to the physical and human problems, the social and economic problems that know no fictitious local boundaries. The capacity to restore the power to plan and govern to jurisdictions which otherwise would remain separated is an expression, under our constitutional system, of the federal residuary legatee.

The Federal Government as an Innovator in Coping with Metropolitan Problems

It is sometimes assumed that the absence of a formal framework for metropolitan affairs means that the metropolis is a decision-making vacuum. In reality, even though they have no government with competent areawide jurisdiction, most metropolitan areas have rich and intricate, bewildering and gaudy if you prefer, patterns of interests and associations. One result of this interaction should be noted: because of the conflict among these galaxies of interests, their momentum normally operates to maintain the status quo. Analysis may indicate a dissatisfaction with the way metropolitan problems are tackled, either because there is an overlap of agencies, or because there are inequities in the provision of public services, or failures to face up to and deal with issues, but such analysis must take into account the existing arrangements and groupings of power. Thus, the first step in bringing about any change in metropolitan affairs must be an awareness of this deep freeze; some melting power must be added to break down the old patterns or, at least, to reconsider their merits *de novo*.

Currently the only source of such heat is the federal government. Any readjustments in local decision making to emphasize metropolitan considerations are abidingly stymied at the local level. The New Jersey Supreme Court, in the *Vickers* case, had to fall back on conventional wisdom and technicalities to avoid bruising confrontations on territoriality. This is the dilemma even for a state court with state jurisdiction, and it provides an *a fortiori* revelation of the impotence of local institutions and authorities when it comes to solving (and this is the key term) the myriad problems associated with metropolitan governance in a nonlocal or nonisolated context. Despite any reluctance the federal government may have to intervene, the locking-in of local processes and the insistence on the old patterns fuels such intervention.

What is described here is a common recurrence where layers of government coexist. Similar interventions have been ascribed to the Warren Supreme Court. Despite the Supreme Court's traditional avoidance of political questions, the total freezing of the political processes on race relations, and later on legislative reapportionment (where those in power in the political arena had most to lose from reapportionment and could block any proposed change), has forced it to intervene. If it is believed that wider urban issues are being neglected, or that areawide considerations are not being taken into account, it is self-defeating to leave our patterns of urbanization to the competing and inconsistent actions of thousands of local government units. It would be highly unrealistic to allocate to local units the responsibility for initiating a metropolitan political accounting system. Why should they initiate one? A degree of national involvement in local affairs becomes essential if extraterritorial needs are ever to enter the equation. To this extent, certainly, Judge Hall's reasoned dissent is vindicated when he calls attention to the undesirable side effects of unfettered local decisions. Yes, there are misallocations and maldistributions requiring redress.

Yet we must be careful that there be no misunderstanding, for this is a subject that bristles with emotion. Although the initiative for change is national, the substantive framework within which such change must take place should be constructed according to local specifications. The components of metropolitan policy need to be shaped by decision making at the local levels; the role of the national government is properly that of catalyst for change, letting the fresh air into the corridors but not constraining its direction and flow. This seems to me to be more than sound philosophy. Congress has recognized this verity in its repeated and insistent requirement that grants be made to conform with local plans that are essentially the products of local people; but these plans, in turn, are then to be made on an areawide basis.[24] Further, the pressures of local groups are such that these edicts will not lie fallow as idle platitudes.

The Federal Government as an Implementor of
Local and Metropolitan Policies

I am not certain how to describe this particular function. In this particular context, it is the merger of thought and deed, of planning and action. As the New Jersey Supreme Court in *Vickers*, so the Congress and the executive branch are faced with realities. Transcendental plans which do not lead to action—good or bad—have begun to leave people restless. One of the main aspects of the federal strategy which we can discern thus far is that of making the metropolitan plan a meaningful instrument for problem solving, or, as President Johnson stated in 1966, "A metropolitan plan is not to be kept in the drawer collecting dust." In short, the stage should be set so that local governments are encouraged to carry their own efforts a big step further toward ensuring that their plans are practical and that they are producing results.

Easier said than done. Too often the planning profession regards its job as completed once it has drafted the imaginative comprehensive plan. And many decision makers would just as soon leave it there—to be hauled out for the League of Women Voters or the travelling inspector from the federal government. A reaction to this futilitarian phase has focused attention on the ability of the public sector to continue the direction and momentum of the metropolitan plans—a continuation which means that attention is given to the questions of adequate funding and also to the relative value judgments that are made in the face of competing opportunities to expend both public works and energies. Before 1966, federal-aid programs included few requirements for the implementation of metropolitan planning. The 701 program spearheaded the major thrust. Then Congress began to add conditions, especially to the HUD grant programs, that the aided project conform to a comprehensive plan

[24] 40 U.S.C. §461(a) (2) (Supp. V, 1970).

drawn up for the entire metropolitan area. Once a comprehensive planning scheme has been formulated, a series of Acts began to assert, the public sector response should move on toward action.

In the reform housing laws of 1966,[25] which launched the Model Cities program, Congressional attention was turned toward externalities of local governments through Title II, the Planned Metropolitan Development Act.[26] Two sections should be noted. They were introduced as competitive programs in the Bill stages and vied with each other through the Committee stage but ultimately, as reluctant bedfellows, found a common haven in the title.

The first of the two is Section 204.[27] Under its provisions, all applications for federal loans or grants to assist in projects that have metropolitan impact—for airports, water supply, highways, transportation facilities, law enforcement facilities, regional parks, and open spaces—must be submitted first for review to a designated areawide agency. This agency is then called on to make comments and recommendations. The agency's report, explaining how the project is consistent with comprehensive planning development for the metropolitan area, is sent to the federal agency to which the funding application is submitted. The comments are "for the sole purpose of assisting it in defining whether the application is in accordance with the provisions of Federal Law which govern the making of loans or grants." This throws the issue back to the original grant statute to see if that law contains a requirement for areawide planning.

Section 204 was intended only to assist the funding agency in its decision making; the evaluations were not to be binding; therefore, few interest groups were moved to object to the enactment of this provision. All the agency is doing in its report is furnishing its judgment that the particular project for which the local government is seeking funds is, or is not, in accord with the metropolitan planning for the area. Yet, the provision introduces a lever of potentially great import. I emphasize "potential," because many planning agencies have not yet undertaken broad studies nor adopted policies which enable them to exercise this power wisely. Either that, or they are too weak or too indecisive at present to pull the lever to the maximum feasible extent. But nevertheless the provision has enormous possibilities. The contents of a plan can affect the distribution of monies. For it is an administrative reality, other things being equal, that a national department will be more reluctant to make a grant if there is some statement on record that the proposal runs counter to the sound development of the area as a whole.

[25] Demonstration Cities and Metropolitan Development Act of 1966, 80 Stat. 1255 (codified in scattered sections of 11, 12, 15, 16, 40, 42, U.S.C.).

[26] 42 U.S.C. § § 3331–39 (Supp. V, 1970).

[27] 42 U.S.C. § 3334 (Supp. V, 1970).

In this sense, the federal government is, it can be said, helping to carry out a local authority mission. It gives local units an opportunity, through their metropolitan representatives, to participate in decisions which were previously closed to them. Where formerly there would have been a "private" agreement between a federal agency and a local agency, now the metropolitan representatives of all the local government units can indicate, for instance, how a proposed capital expenditure by one unit can thrust unfair externalities on the others or how a particular plan will fail to further the overall development strategy of the entire area.

The second important feature introduced in 1966 was Section 205.[28] A counterpoint to the provision for supplementary grants to model cities, this section provides a 20 percent supplement, by way of an override, to basic federal grants designated for areawide development projects.

The grants supplement existing federal assistance to programs for transportation facilities (transit, primary and secondary highways, and airports), water and sewer facilities, recreation and other open space areas, historic preservation, libraries, and hospital and medical facilities. Grants may be made for such projects in metropolitan areas only if certain requirements are met for metropolitan planning, programming, and local implementation of areawide plans and programs.[29]

When the individual applicant is a public body, it must show that public facility projects and other land-development activities over which it has jurisdiction, and which are of interjurisdictional or areawide significance, are being carried out in accord with the required metropolitan planning and programming.

Where the applicant is a unit of general local government (county, city, town, or other general-purpose political subdivision), it must show that it is assisting in, and conforming to, metropolitan planning and programming through (1) the location and scheduling of public facility projects (whether or not federally assisted), and (2) the establishment and consistent administration of zoning codes, subdivision regulations, and similar land-use and density controls.

[28] 42 U.S.C. § 3335 (Supp. V, 1970).

[29] To establish eligibility for supplementary grants in a particular metropolitan area, projects must show that (1) comprehensive planning and programming on a metropolitan scale provide an adequate basis for evaluating the location, financing, and scheduling of public facilities and land development (whether or not federally assisted) of metropolitanwide or interjurisdictional significance; (2) adequate metropolitanwide institutional or other arrangements exist to coordinate local public development policies and activities affecting the development of the area; and (3) public facility projects and other land developments (public or private) which have a major impact on the development of the area, are in fact being carried out in accord with metropolitanwide comprehensive planning and programming.

Where the applicant is a special-purpose unit of government (special district, public-purpose corporation, or other limited-purpose political subdivision, excluding school districts), then both it and the unit of general local government with jurisdiction over the location of the project must qualify under the program.

Supplementary grants under Section 205 are intended to go to states and localities that are achieving a high order of efficiency and economy in constructing federal-aided public works. Here again, therefore, there is the potential for carrying out the thrust of the 702 program. The real significance of the program is its leverage power to foster the kind of growth that maximizes and protects capital outlays. (Due to pressures of economy, however, nothing has yet been funded, nor is the present administration apparently willing to fight for such appropriations in the face of what it deems the inflationary tide confronting it.)

Naturally, as an incentive (and supplementary) device, Section 205 goes much further in setting out metropolitan criteria than does the requirement for comprehensive planning and programming under an ordinary 702 grant. Under Section 205 the Secretary must be satisfied that the comprehensive planning and programming provide an adequate basis for evaluating the location, financing, and scheduling of individual projects that have areawide or interjurisdictional significance. As a precondition, there must be areawide institutional arrangements that are adequate to coordinate local public policies and activities affecting the development of the area. It must also be demonstrated that public facility projects and other land developments are in fact being carried out in accordance with areawide comprehensive planning.

In addition, if a project is to qualify for a supplementary grant, it must be linked in with private development; that is, there must be evidence that the administration of subdivision codes, zoning codes, and land-use and density controls has been consistent with the metropolitan plan, so that the objectives of the plan are also carried out in the regulation of private actions, which are so important for the use and development of urban land.

As the draftsman for Section 205, it is only proper for me to point out a weakness in its construction, though not, I believe, in its architecture. It may have gone too far too fast. Its potential strength boomeranged: it would have provided a source of monies to metropolitan areas, essentially in the form of bloc grants, at a time when we were still feeling experimental about bloc grants and tentative about our metropolitan psychology. It may also have placed too great a strain on the credibility of the planning profession. By rewarding and putting forth cash incentives, planned projects became realities and their validity and claims could be put to the test. Sometimes, especially for a profession dealing with so many uncertainties as it peers into the future, it is a mixed blessing to be taken too seriously. But Section 205 still stands as the high-water mark: local capital budgets and the private regulatory activities affecting the development of the area as a whole are now required to tie into a metropolitan development plan with clearly stated assumptions, programs, and goals, and these, in turn, can be tested by parties, both governmental and private, and handled in the traditional state judicial forum.

Still other issues arise, as one examines Section 205 more deeply. Although such a program, which deals seriously with urban issues on an area-wide basis, coincides with the true interests of the federal government, its proper niche in the Washington firmament is difficult to find. The 1966 Act placed the program in HUD—the agency already administering the Sections 701 and 702 programs. But HUD is simply one department. Others (Transportation; Health, Education, and Welfare; Labor) also give monies for planning and for local government activities which have areawide impacts. Through Section 205, however, HUD can make available supplemental grants to those areas in which each unit is ready to regard itself as part of a functioning whole and is ready to work with others on a cooperative and coordinated basis. HUD thus assumes a role on the national level with respect to these operations that the metropolitan agency tends to assume on the local level: it does not have a line function; it does not have grant money to undertake a program; it does not even have final responsibility for the successful conception and execution of a project. Instead, what it does have is a kind of kibitzing power, a vague, rather ill-defined comprehensive function. Unless some clearer mandate for coordination evolves, responsibility for the program might be better moved to the office of Management and Budget or some other White House center.

Section 205 provides HUD with this comprehensive function, eleborated as best it can be, and this has happened despite an inherent philosophic cleavage in Washington. The Congress has fully accepted, and is apparently ready to continue, the areawide requirements—but only within certain practical restraints. There is inevitable institutional reluctance to accept procedures smacking of "a fourth power." To the extent that rational planning is introduced into the system, to the extent that elitism or technocratic knowledge emerges as a reason for giving or not giving grants, there is bound to be competition with the political process, which the Congressman exemplifies. Although he recognizes the need for metropolitan efficiencies and he recognizes the need to deal with external ramifications, the Congressman is confronted with the far greater need to be reelected. To him, a project proposal represents the winning of a grant for his district, proof that he has done something recently for his constituents. And so far as his immediate needs are concerned, therefore, areawide requirements become annoying technicalities.

This was driven home to me during a rezoning case. On its last night in office, the lame-duck Montgomery County (Maryland) Council had a final fling; it amended its zoning ordinance to permit garden apartments and other intensive land-use developments throughout the county. Such developments would have had the effect of encouraging building on land formerly designated as open space and as potential reserve water supplies. Still pending at HUD (not surprisingly, in view of the snail-like speed of bureaucratic action) was an earlier application from the county for open-space grants based (under the congressional mandate) on an areawide plan for recreation and parks. This plan, the firm underpinning, would have been totally subverted by the rezoning. HUD, therefore, suspended the processing on the open-space applications. To the Congressman from the area, this meant one thing, and one thing alone—that his home district was not receiving a grant. He clamored loudly. Small bother that the basis for the open space plan (which, incidentally, had been approved by President Kennedy as part of the Washington Metropolitan Area Year 2000 Plan), and against which the project had statutorily to be measured, was being eroded.

This tug of powers has been illustrated elsewhere, again and again, where one unit of government within an area has been ready to cooperate but another has not. Congressmen have constantly been puzzled that there should be formalistic insistence on areawide cooperation before a grant to their constituents can be made.

But while I point out these difficulties with Section 205, I still remain a believer. The elaboration of experience becomes essential. Section 205 should not be laid aside simply because it lacks the precision of older and more established operations. The interrelations of activities and their external by-products are coming increasingly to public attention. This we can see in the current public concern over the growing inroads into our environment and over the by-products of transportation and housing as they affect other uses and developments. Comprehensive planning can give us a mechanism for calculating more accurately the costs to the local government. It also gives a social accounting over the whole metropolitan area, so that we can arrange for social compensation for the costs of individual action. Funding under Section 205 can convert the metropolitan development plan from kibitzing to catalyst.

In an important way, then, Section 205 stands as a watershed for metropolitan development in terms of federal concern. The Section went through the authorization stage twice but failed to gain appropriations. Ideas and programs do not come overnight. Nor are they accepted as soon as they are enunciated, not even when they are clarified. They need to be examined, explained, discussed, pushed. In a democratic society, politics is the art of finding a consensus and a balance among different interests.

The injection of the doers and the realists into the metropolitan planning process makes it that much more difficult for the metropolitan planner; but this injection also makes his job far more interesting. If public projects are executed and the control of private land uses is handled in accord with a metropolitan plan, this means that the plan has received validity in the real world. It has bite. It has meaning. It gets people upset—unfortunately. But when a plan makes people uncomfortable—from time to time, in proportion, and on its merits—it can be said to have impact and reality in an otherwise gossamer world.

If supplemental grants under Section 205 are given only where public capital expenditures and control over private development are being carried out in accord with the locally adopted areawide plan, then the plan has substantive impact on the operations of the land market and on the real world economy. The metropolitan plan will not be idle chatter, but something in which men's lives, men's energies, and men's money are directly affected. And, hopefully, professional planners and local governments can then begin to take their guidance work seriously. The courts, too, can begin utilizing the plans in a way that Judge Hall would have liked to do had they been available.

The Federal Government as Sponsor of National Policies That Require Metropolitan Cooperation

The nation as a whole has adopted policies to cope more effectively with the problems of the cities that have crucial metropolitan implications.

Various constitutional and ethical considerations come to the forefront, which are the special province of the national government. There is, for example, the overriding objective of "integration," the reduction of race, religion, and ethnic origin to the status of irrelevancy in the operations of the public sector within the metropolitan area. The federal role in pinpointing the transcending values that we as a society seek to foster when we tailor institutional structures is carried out at various times, and with varying emphases, by the different branches—through the moral leadership of a presidential message, the due process clarification of a federal court, the constitutional condition attached to a Congressional appropriation.

Among the metropolitan factors surely in need of reexamination on this score—candidates for paradigm alteration—are the attitudinal and institutional models now employed for the delivery of public services. In the brief in *Hawkins* v. *Town of Shaw*,[30] Professor Fessler and I characterized these areas as that functional plateau at which the many must depend upon the stewardship of a few for the rendition of those services that cement a collective society. We cannot behave responsibly when we entertain the possibility of

---

[30]No. 29013, Fifth Circuit Court of Appeals, Jan. 28, 1971.

reformation if we do not evaluate the equalities, or inequalities, that would follow any changes in the system of demand or in the models of supply.

An outstanding example is in the area of housing. In the Housing and Urban Development Act of 1968,[31] the Congress chose to establish a quantitative measure of its target, which was to meet the needs it had enunciated in the Housing Act of 1949—"a decent home in a suitable living environment for every American family."[32] Spelled out, this quantitative measure was for the construction of 26 million units over the next 10 years (of which fully 6 million were to be subsidized) and also the rehabilitation of the substandard existing stock. Such an ambitious national program is already stumbling over local hurdles, however. Exclusionary zoning ordinances have prevented low- and moderate-cost housing from entering into suburban enclaves where otherwise it would have spread through forces of the private market. If we are to achieve a federal program for subsidized housing throughout the urban centers, it becomes necessary to prevent local communities from raising obstructions through the use of local land-use controls or other tactics that reinforce the status quo. In practical terms, HUD's broad housing objectives require the dispersion of many of the poor throughout the region, and this can be attained only by a metropolitanwide policy. Indeed, the Kaiser Commission on Housing even went so far as to suggest that the Secretary of HUD should have the power to preempt local zoning codes so that they would not apply to federally subsidized low- or moderate-income housing projects.

Environment has become another crucial policy area. In his State of the Union Message of 1970,[33] President Nixon outlined a strong national program to control pollution and to try to " . . . pay our debt back to nature." A community that pours the wastes from its industrial plants or its municipal sewage into a river which crosses state lines provides a good illustration of the types of problems produced by environmental factors. Again, air pollution is not confined to state boxes. Aside from the institutional vacuum, which requires federal intervention in its role as a steward of the nation's land resources, such boundary dissonances render wasteful other expenditures made from the federal fisc.

Other examples of national policy that require metropolitan coordination come readily to mind. The redistribution effect of the national unit within the extended boundaries of the country may be frustrated by the preoccupations of a state or local unit of government. Grants under the Law Enforcement Assistance Administration (LEAA) and Title I of the Higher Education

[31] 82 Stat. 476 (codified in scattered sections of 5, 12, 15, 18, 20, 31, 38, 40, 42 U.S.C.).

[32] 42 U.S.C. § 1441 (1958), *as amended*, 42 U.S.C. § 1441 (Supp. V, 1970).

[33] H.R. Doc. No. 91–226, 91st Cong., 2d Sess. (1970).

Act are recent examples of this hazard. Falling far short of the current proposal for a national growth policy (which would require coercions and, more probably, inducements by Washington on an unprecedented scale), many of the redistributive and economic growth programs already adopted are strongly affected by the existence of the many and uncoordinated local sovereignties.

The Federal Government as a Clearing House and Source of Technical Assistance

In another age, a great spokesman for creative change, Mr. Justice Brandeis sustained a state law that went far toward imposing economic regulation on the industries within its borders and wrote of the "laboratory of the states." The independent states were set up, so the theory ran, in such a way that each could hammer out its own destiny and could learn from the successes and mistakes of the other. Increasingly, however, with the emergence of national corporations and the evolution of a national home mortgage market and a national building industry, a state can take leadership steps only at a sharp risk to its own immediate welfare. Furthermore, the expense of many of the new technological devices makes it prohibitive for any one state to engage in such innovation. A new type of electric bus, to take a small example, is too expensive for any one agency to order; but a large-scale purchase, underwritten by the federal government, makes such revision feasible.

In addition to this wholesaling aspect, the federal government, by its ability to utilize professional specialists, can bring the latest thinking and experience to bear on our perennial metropolitan problems. A history of change in the components of the 701 metropolitan plan is illustrative. In a period of 2 years, guidelines and techniques were evolved for dealing with matters such as public transportation on a large scale, housing on a metropolitan area basis, regional urban design projects, the administration of criminal justice as a function of chief executives in the metropolitan area, the stirrings of program planning and budgeting systems (PPBS) in metropolitan budgeting and the use of metropolitan indicators to quantify goals; rather than stressing traditional physical and land aspects, there emerged a federal thrust toward dealing with metropolitan issues of health, education, and low-income groups.

Such federal leadership to local governments is reinforced by its function as a clearinghouse, by its ability to see which successful ideas are susceptible to general applicability and transferability. This comparative approach also gives a sense of balance and proportion: in so new an area of concern, there can be federal leadership and support for many of the hard-pressed states and local governments.

The Next Step

Thus far I have tried to delineate the rationale for federal involvement in, and responsibility for, metropolitanism on the lines of what appear to be the political realities. They articulate what I believe has come to be a consensus, and a minimal engagement of federal resources. Within this framework, of course, there lie ranges of federal action and potentialities, depending on the degree of seriousness with which the metropolitan problem is perceived by the federal establishment. Over and above the federal government acting as banker, carrot and stick, and institution builder, there are the powerful substantive levers of the national government which can help solve metropolitan urban problems redefined as national in scope. New communities offer a prominent example of this: if one is serious about a national strategy of population dispersal and of the optimum patterns of investment and population settlement in the nation, the role of the federal government becomes crucial. With the levers provided by this kind of assistance or by the proposed Urban Development Bank,[34] what we have described above as the implementation of metropolitan plans becomes a far more feasible proposition. And still adhering to the framework of the division of powers, a vigorous enforcement of land reserve policies and land banking on the national level can make reality out of metropolitan plans that would break the white noose around black central cities and increase the availability of low-income housing to the entire nation. The issues of metropolitanism, by their very nature, spill over into other fundamental concerns of a nation. Overarching all the individual tactics and the sundry grants-in-aid, all of the incentive, sanction, and regulatory devices that the American ingenuity can invent, is the necessity of a moral philosophy, an urban ethic which can provide a light and an insight to the technician, whether he be institutional doctor, reformer, or hard-headed financial advisor. The painfully hewed-out measures for effective metropolitan governance require constant attention. In their application to the unforeseen contingency, there is need for an ethic on which to base them. Here lies a basic challenge and an opportunity for political leadership.

In many ways, and perhaps too often for the amount of limited capital that it possesses, the moral leadership that must precede political action has come from the judicial branch of government. Judge Hall writes of the underlying philosophy applicable to the developing local governments within a vast metropolitan complex: the general welfare should transcend "the artificial limits of political subdivisions and cannot embrace merely narrow local

---

[34]See Charles M. Haar and Peter A. Lewis, "Where Shall the Money Come From?" *The Public Interest* 101 (Winter 1970), for an amplification of this proposal.

desires." "Deeper considerations," he said, "intrinsic in a free society gain the ascendency and courts must not be hesitant to strike down purely selfish and undemocratic enactments." The *Vickers* dissent concludes with the proper pragmatic challenge to planners and administrators: "To reiterate, all the legitimate aspects of a desirable and balanced community can be realized by proper placing and regulation of uses, as the zoning statute contemplates, without destroying the higher value of the privilege of democratic living."

The governance of metropolitan areas raises grand issues of constitution making and of the roles of governments and their relation to the individual. Earlier, and speaking for the national judiciary, this idea had been adumbrated in its broadest context by Mr. Justice Cardozo. The Constitution, he said, " . . . was framed upon the theory that in the long run, prosperity and salvation are in union and not in division." In metropolitanism, then, as it evolves in the 1970's, lies the federal government's opportunity to exercise its great potential leadership toward this union, first, for salvation, and next, for prosperity.

For Product Safety Concerns and Information please contact our
EU representative GPSR@taylorandfrancis.com Taylor & Francis
Verlag GmbH, Kaufingerstraße 24, 80331 München, Germany